Fossils
The essential guide

Paul D. Taylor

Published by the Natural History Museum, London

First published by the Natural History Museum,
Cromwell Road, London SW7 5BD

This edition published in 2025

© Trustees of the Natural History Museum, London, 2025

ISBN 9780565095635

The Author has asserted his rights to be identified as the Author of this work under the Copyright, Designs and Patents Act 1988.

All rights reserved. No part of this publication may be transmitted in any form or by any means without prior permission from the British publisher.

A catalogue record for this book is available from the British Library.

Designed by Mercer Design, London

Reproduction by Saxon Digital Services
Printed in China by Toppan Leefung Limited

Front cover: Sectioned fossil of a nautiloid from the Middle Jurassic of Somerset, UK.

Contents

	PREFACE	04
1	FOSSILS, ROCKS AND GEOLOGICAL TIME	06
2	THE OLDEST FOSSILS	28
3	TRILOBITES AND OTHER ARTHROPODS	44
4	AMMONITES AND THEIR RELATIVES	62
5	SUPERABUNDANT SHELLS	76
6	INVERTEBRATES OF ALL SHAPES AND SIZES	102
7	THE PHYLUM OF FIVE: ECHINODERMS	126
8	FISHES AND AMPHIBIANS: FROM FINS TO FEET	146
9	REPTILES AND BIRDS: THE RISE OF THE ARCHOSAURS	162
10	OUR FAMILY AND OTHER MAMMALS	182
11	FOSSIL PLANTS: THE GREENS JOIN THE PARTY	204
12	MICROFOSSILS BY THE MILLIONS	222
13	TRACE FOSSILS: SNAPSHOTS OF ANCIENT BEHAVIOUR	236
	FURTHER INFORMATION	250
	INDEX	250
	ACKNOWLEDGEMENTS	256
	PICTURE CREDITS	256

Preface

The possibility of the existence of life beyond our planet has long fascinated humanity. One of the objectives of unmanned missions to Mars has been to find indications of living organisms, either still alive or long since dead and represented by fossils in the rocks of the red planet. While few scientists would bet on the existence of living organisms on Mars today, a significant number would probably gamble on life having once existed there. After all, there are numerous features on the Martian landscape pointing to the former presence of liquid water, believed to be a prerequisite for life as we know it. Perhaps one day in the distant future all life on our own planet will have vanished, leaving only fossils to show that Earth was once populated by a bewildering variety of living organisms. Whether or not fossil evidence of life is ever found on Mars, the rich fossil record of ancient life on our own planet is a remarkable legacy to be cherished.

Just as the novelist L.P. Hartley characterized the past in human terms as a foreign country, so the ancient past in evolutionary terms can seem like an entirely alien world, the more so the further back we go in time. It is the job of the palaeontologist to reveal the mysteries of the organisms that have lived on planet Earth since its formation some 4,600 million years ago. And fossils are key to this quest. Without fossils, would the idea of evolution ever have taken hold in the nineteenth century? Probably not. Fossils are not crucial in working out the details of evolutionary change, but they do show us that animals and plants very different from any living today once populated the Earth. Some fossils provide us with links between living organisms that bear no obvious relationship. Every single fossil has multiple stories to tell. Stories of birth, growth and death. Stories of evolution and adaptation to life in different environments. Stories of species originations, persistence and extinction.

Global environmental change is an increasing concern for humanity. Sea level is rising, temperatures are climbing, extreme weather is becoming more frequent, the oceans are acidifying and large numbers of species are becoming extinct or are facing extinction. Although contemporary changes appear to be happening at an unprecedented rate, with global warming now about 50 times the rate of the previous 21,000 years, our planet has been in a restless state throughout its 4,600 million-year existence. The evolution of life on Earth has been enacted out in an ever-changing environmental theatre. Fossils not only play an important part in tracking ancient environmental changes, but also provide insights into the consequences of these changes on living organisms.

Fossils also have practical value in correlating distantly separated outcrops of rock that formed at the same time. This has been essential to geologists in search of mineral resources when attempting to understand geological structures from incomplete surface information.

There can be few better introductions to science for children than fossils. All sciences at some point involve three activities: discovery, naming and interpretation. In the case of fossils, children experience great delight in finding a fossil and holding in their hands the remains of a creature that lived millions of years ago. The next stage is to identify the fossil. What kind of animal or plant was it? Is it possible to give it a genus or species name? Then comes interpretation which, like naming, can be undertaken at whatever level is appropriate. How and why was the fossil preserved? How did the organism live and when? What can it tell us about the ancient environment in which it lived and about evolution?

This book sets out to provide a broad introduction to fossils and is aimed at the interested layperson. After a chapter explaining how fossils are formed and some basic aspects of geology relevant to the fossil record, the chapters that follow concern biological groups that are well-represented in the fossil record. These are the fossils most likely to be encountered, not just in museums but also wherever fossil-bearing rocks are exposed on the land surface, as well as in the stones we use for building. Of course, in a short book like this it is impossible to do more than skim the surface of the vast field of palaeontology, and the reader is directed to the numerous specialist websites and books (see p. 250).

1 Fossils, rocks and geological time

Mammoths, dinosaurs, megalodon sharks, ammonites, clubmoss trees – these are all extinct organisms known only from their fossilized remains. Hundreds of thousands of different fossil species have been found in the rocks of our planet. They paint a vibrant, albeit incomplete, picture of the changing inhabitants of Earth stretching back more than 3,500 million years that, without them, would be a blank canvas, or at best a sketchy rendition. Everyone has heard of fossils but not so many are able to define them. Most people have two ideas: that they are extremely old objects, usually of biological origin, and that they are contained in rocks dug out of the ground. The first is reflected in the use of the term fossil for things that are ancient, outdated or obsolete, the second explains why palaeontologists are often confused with archaeologists – both dig up material for study. However, whereas archaeologists deal with artefacts of human culture, palaeontologists are interested in the remains of ancient living creatures preserved in the rocks. Before exploring the rich diversity of fossils, it is worth dealing with some key questions.

What is a fossil?

The word fossil comes from the Latin *fossus*, a ditch, and centuries ago it was applied to literally anything dug out of the ground – including fragments of pottery and minerals, as well as objects that appeared to have an organic origin – at a time before the origin of fossils was fully appreciated; these were often referred to as 'formed stones' by naturalists. Formed stones are what we would now call fossils – the remains of ancient living organisms preserved in rocks through natural processes of burial. Unsurprisingly, fossils most often consist of those parts of organisms that were resistant to decay. Shells, bones and teeth are the typical stuff of fossils, whereas muscles, nerves and veins are seldom preserved. There are, however, some notable and spectacular exceptions, as we will see later in this book. Another kind of fossil, known as a 'trace fossil', results from the activities of living organisms leaving footprints, burrows etc. (see Chapter 13).

OPPOSITE Opalized shell of the Cretaceous gastropod *Euspira* from White Cliffs in New South Wales, Australia. 2.5 cm (1 in) in size.

When does something become a fossil?

It would be nonsensical to label as a fossil the shell of a mussel that died yesterday and was buried on a beach by sand from the evening's incoming tide. There is an obvious grey area between this scenario and an arbitrary minimum age of 10,000 years, which is often used in the definition of a fossil – roughly the age of the boundary between the Holocene and older Pleistocene epochs of geological time (see p. 19). Younger organic remains from Holocene deposits may be referred to as 'subfossils'.

How are fossils formed?

The journey from a living organism to a fossil can be complex and varied. Perhaps the simplest sequence would be:

- an animal with a biomineralized shell dies on the seafloor
- the muscles and other soft parts are consumed by scavengers or rot away
- sand or mud, perhaps swept in by a storm, buries the shell
- as further layers of sediment are deposited, the shell becomes ever more deeply buried and beyond the reach of erosion that might return it to the surface
- the sediment hardens into a rock, a process known as 'lithification', due to compaction caused by the weight of overlying layers, together with the growth of tiny crystals of inorganic cement that bind the sediment grains tightly together
- any residual organic material within the shell disappears, and changes may also occur in the mineral component of the shell
- the fossil shell is brought to the surface again by natural erosion or human quarrying activities.

There are numerous variations on this general pattern. For example, organic remains can be transported by currents and buried a considerable distance from where the animal or plant originally lived. In animals with skeletons consisting of multiple elements, such as the bones of vertebrates, the different parts can become dispersed before burial, with smaller or less-resistant elements not surviving the attention of scavengers or being lost through erosion.

When living animals are buried, their soft parts normally decompose in the sediment, but very occasionally muscles and other organs are replaced or overgrown by minerals before they have decayed. Deposits containing such exceptionally preserved fossils are known by the German term Konservat-Lagerstätten. They include the famous Cambrian Burgess Shale (p. 39) and Jurassic Solnhofen Limestone (p. 58). Mineralization of organic tissues in these deposits is usually due to the presence of bacteria: sulphate-reducing bacteria may cause the precipitation of iron pyrites (so-called fool's gold: FeS_2), while other kinds of microbes can expedite phosphatization of fossils in low-oxygen environments where plentiful organic matter provides a source of phosphorus.

Once fossilized, shells and bones are invariably heavier than those of living animals. This is because the spaces once occupied by soft tissues are often filled by minerals precipitated from fluids passing through the rock – formerly porous bones become solid structures, and delicate shells are reinforced by these additions. On other occasions, shells and bones are lost through dissolution, leaving holes in the rock. These holes represent natural moulds. In the case of shells filled with hardened sediment, the sediment core is an internal mould, often referred to using the German name steinkern translated as 'stone nut', whereas the impression of the shell in the surrounding sediment is an external mould. Both are negative impressions of the surfaces of the fossil. Moulds can be infilled subsequently by mineral growths, leaving a natural cast of the fossil. The most striking examples of this process are the opalized fossils from the Cretaceous rocks of South Australia. Opal, an iridescent, hydrated form of silica, fills voids in the rock, in this case those left by dissolution of fossil shells, bones and teeth. Clearly, the original internal microstructure of fossils preserved as casts is completely lost, but in another kind of replacement preservation even these details may survive, albeit usually in a degraded form. This is when the minerals of the fossil are progressively substituted along a thin fluid film by different minerals, a process called neomorphism. For example, mollusc shells originally made of aragonite can be replaced by the more stable mineral calcite conserving ghosts of the aragonite crystal structure.

The fossilization of plants lacking biomineralized skeletons can be somewhat different from the processes outlined above and is discussed separately in Chapter 11.

LEFT Internal mould, known as a 'steinkern', of the echinoid *Clypeaster*. From the Miocene of Spain, the fossil is 11 cm (4¼ in) in diameter.

FINDING FOSSILS

ABOVE The large Kotouč Quarry at Štramberk in Czechia exposes the fossil-rich Štramberk Limestone of latest Jurassic and earliest Cretaceous age.

It is not difficult finding places to observe or to collect fossils. Fossil collecting today continues a practice that has been taking place for thousands of years: for example, fossils were picked up by prehistoric humans and used in jewellery and even placed ceremonially in graves. Indeed, evidence from archaeological sites shows that not only modern humans collected fossils but also earlier species of our genus such as *Homo heidelbergensis* and *Homo neanderthalensis*.

Sedimentary rocks containing fossils are visible in natural outcrops, including sea and river cliffs and deserts where soil and plant cover is lacking, as well as in man-made exposures such as quarries, road-cuttings and temporary excavations for building foundations. Before venturing into these places, it is essential to ascertain whether they are protected sites, and to obtain permission from landowners if this is required. It is worth noting that in some countries (e.g. Australia, Brazil, Greece) collecting fossils and exporting them requires a permit or is banned entirely. The risks from falling rocks, mudflows, incoming tides, slippery rocks, etc. should be assessed whenever undertaking fossil collecting. Safety helmets and high visibility vests are mandatory in some places, and it is never advisable to embark on collecting trips alone in case of an accident.

Having an appropriate search image is important when looking for fossils – in the case of ammonites, for example, this would be for spiral patterns in the rock. Picking up fossils from weathered material on the ground is often more productive than focusing attention on fresh rock faces, where fossils may be less clearly visible and not easily extracted. A small hand lens, sometimes called a jeweller's loop, with a magnification of between ×10 and ×20 is needed when searching for small fossils. Collecting fossils from hard rocks requires a geological hammer and a chisel, while

ABOVE Roadcutting near St George in Utah, USA exposing fossil-bearing rocks of the Triassic Virgin Limestone.

LEFT Cretaceous Chalk forms Abbot's Cliff near Dover in Kent, England where it is possible to collect fossils from the foreshore.

trowels and sieves are more appropriate tools when collecting from loosely cemented rocks. Fossils should be wrapped in the field – paper towels and kitchen foil are useful for this purpose – and placed in labelled plastic bags. Detailed preparation of specimens can be undertaken later using a mounted needle or small chisel to remove the surrounding rock matrix, before final cleaning with a soft brush such as an old toothbrush or an ultrasonic cleaner.

Numerous museums in Britain and abroad have excellent displays of fossils. Regional museums tend to exhibit locally found fossils, whereas larger national museums often feature fossils collected from further afield, including the rarer and more spectacular fossils, such as the ever-popular dinosaurs. However, there is an alternative way of seeing fossils in urban settings. It may surprise many how easy it can be to observe fossils in building stones in cities. 'Pavement palaeontology', as it has been dubbed, offers the possibility of discovering fossils in unexpected places wherever sedimentary rocks – particularly limestones – have been employed for construction. Old churches and other buildings, as well as walls, exterior pavements and interior stone floors, can be excellent places to spot fossils. These are often seen in section in cut blocks of stone. For example, fossils of Jurassic ammonites, belemnites and sponges are visible in the numerous modern buildings using 'Jura Marble', quarried in southern Germany, as a cladding material or as polished floor tiles. Historic buildings can also be replete with fossils. Hampton Court Palace in west London boasts paving of Ordovician Öland Limestone from Sweden full of orthocone nautiloids (p. 72), columns of Jurassic Portland Stone containing bivalve and gastropod fossils, cobblestones of Cretaceous Purbeck Marble packed with mollusc shells and even the mounted antlers of a Pleistocene giant deer.

How do fossils get their names?

Fossils mostly belong to major taxonomic groups with familiar popular names – molluscs, dinosaurs, conifers – but very few fossil species have vernacular names. Instead, they must be referred to using a formal scientific name, which can be challenging to the non-specialist and difficult to pronounce. As with modern organisms, the formal names of fossils use the binomial system introduced in 1735 by the Swedish naturalist Carl Linnaeus, in which a genus name is followed by a species name. Both are by convention always printed in italics, the genus name with a capital first letter and the species name with a lower case first letter, as in the dinosaur *Tyrannosaurus rex*, one of very few fossils with a widely recognized formal scientific name.

The names of fossil genera and species are Latinized versions of words derived from a vareity of sources. Some refer to a conspicuous feature of the fossil (e.g. the very large snail *Campanile giganteum*), others are derived from a place name (e.g. the plant genus *Caytonia* from Cayton Bay in Yorkshire, UK), or from a mythological figure (e.g. the ammonite genus *Amaltheus*), or are given to honour a notable person (e.g. the fish *Materpiscis attenboroughi* after the broadcaster Sir David Attenborough). For formal names of species to be valid, they must be published according to a set of rules. A defining type specimen (known as a 'holotype') has to be chosen and placed in a recognized museum where it can be available for restudy, and reasons should be stated why the new species is different from existing species. In its fullest form, the name is followed by those of the authors and the date of publication. For the famous dinosaur mentioned above, this would be *Tyrannosaurus rex* Osborn, 1905 because the dinosaur was first named by Henry Fairfield Osborn in a 1905 issue of the scientific journal *Bulletin of the American Museum of Natural History*. Quite often, the genus name is abbreviated to a single letter: *T. rex* has become an almost ubiquitous moniker for *Tyrannosaurus rex*, which is the 'type species' of the genus *Tyrannosaurus*. If any other species is to be placed in the same genus, it must be shown to share features implying a close relationship to *T. rex*.

One final point – the author and date of a species are often written enclosed in brackets. This signifies that the species has been transferred to the current genus having been originally placed in a different genus. For example, the Cambrian trilobite species now called *Elrathia kingii* (Meek, 1870) was originally named *Conocoryphe kingii* by Meek in 1870. However, in 1915 it was reassigned to a new genus called *Elrathia* because of significant differences from the type species of *Conocoryphe*.

Which types of rocks contain fossils?

A basic knowledge of rocks is essential in palaeontology, if only to know which rocks are most likely to host fossils. There are three main types of rock – igneous, sedimentary and metamorphic. Igneous rocks are formed by the cooling and crystallization of molten magma, either deep within the Earth or as lava cooling on the surface – examples include granite, basalt and obsidian. Sedimentary rocks are formed by the accumulation of eroded fragments of pre-existing rocks carried through the air or in water to a site of deposition – examples include sandstone, shale and limestone; metamorphic rocks are formed by the actions of intense pressure and/or high temperatures on igneous or sedimentary rocks – examples are marble, slate and gneiss.

Fossils have been found in each of the three main rock types, but the great majority come from sedimentary rocks, far fewer are found in metamorphic rocks, and they are exceedingly rare in igneous rocks. Fortunately, sedimentary rocks cover large tracts of our planet and consequently fossils can be found in numerous places.

All sedimentary rocks are deposited in layers called strata, which are separated by bedding planes marking pauses in deposition or changes in the type of sediment deposited. Gravity normally ensures that strata are more-or-less horizontally disposed, although subsequent earth movements may tilt or fold them. An exception to the initial horizontal disposition is when sediments are deposited under the influence of currents on the steep faces of advancing dunes in deserts or underwater as ripples and sandwaves. This kind of deposition results in cross-bedding, in which bedding planes are inclined at an angle to the horizontal.

There are many varieties of sedimentary rocks, some more likely than others to contain fossils. The quality of the preservation of these fossils also varies in different rocks. Sedimentary rocks are classified and named according to the size of the sediment particles of which they are composed, as well as the chemical and physical characteristics of these clasts or grains and how tightly they are bound together. A primary distinction is between siliciclastic rocks and limestones.

Siliciclastic sedimentary rocks are composed of pieces of silicate minerals such as quartz, feldspars and clay minerals. When these pieces are the size of pebbles, cobbles or even boulders, the rock is called conglomerate or breccia, the former if the clasts are rounded and the latter if they are angular. Coarse sedimentary rocks tend to indicate deposition in energetic, erosive environments and do not often contain decent fossils. Sandstones result from the cementation of sand-sized grains. Although fossils may be present, the high porosity of sandstones may allow the flow of acidic pore waters to dissolve any fossils present. Finer-grained siliciclastic rocks formed from mud include clay (uncemented), mudstone (cemented clay or silt) and shale

PSEUDOFOSSILS AND FAKE FOSSILS

Non-biological structures can easily be mistaken for fossils. Referred to as pseudofossils, the most common are concretions and nodules found in sedimentary rocks. Formed by the localized growth of mineral cements (such as calcium or iron carbonate) within buried sediment, concretions vary in size and shape but are typically a few tens of centimetres in diameter and range from perfectly spherical to lens-shaped or irregular and lumpy. It is easy to be deceived into believing that concretions must have an organic origin, particularly those with symmetrical shapes or which by pure chance resemble animals. Take so-called septarian concretions. These contain mineral-filled cracks that have a polygonal pattern on their surface, making them look like fossil turtle shells. Nodules of flint from the Cretaceous Chalk can also be pseudofossils as they frequently take on shapes mimicking organic structures – ducks' heads, miniature human legs, etc. The sheer abundance of flint nodules of differing shapes weathering out of the Chalk and concentrated on beaches and over ploughed fields guarantees that some will by chance resemble fossils of organic origin. Manganese dendrites are another kind of pseudofossil.

These inorganic precipitates of manganese oxides trace fractal patterns along bedding planes and other cracks in the rock, recalling the branching patterns of plants.

The media love a story of deception and can become particularly engaged by tales of fossil fakery. The disproportionate publicity given to fake fossils can lead to the false impression that fakes outnumber true fossils. Nothing can be further from the truth, even though increasing numbers of fake fossils are appearing on the market as the commercial value of fossils increases. Although some fakes are made in attempts to deceive scientists, the majority are manufactured for sale to members of the public. They include whimsical 'fossils' of tiny angels unlikely to fool anyone with a basic knowledge of science, as well as fake trilobites and reptiles which are more convincing as true fossils at first glance. Close inspection, however, will usually reveal evidence that the fake fossil has been carved into the rock, or made from a mould filled with powdered rock plus resin or another kind of cement.

RIGHT Fake fossil of a trilobite 10 cm (4 in) in length.

BELOW A flint nodule by chance resembling the skull of an animal. This 'pseudofossil' was photographed on Epsom Downs in Surrey, England and measured about 25 cm (9¾ in) long.

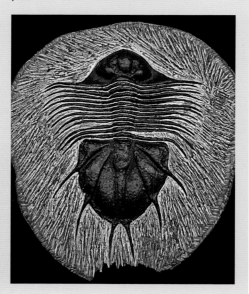

(fissile mudstone). As these mudrocks are less permeable than sandstones, the fossils they contain tend to survive better. Furthermore, the typically tranquil environments of deposition of mudrocks facilitates high-quality preservation. Many of the best-preserved fossils come from mudrocks, especially those deposited in low-oxygen conditions where decay was retarded and there were no large burrowing animals to churn up the sediment and disrupt the nascent fossils. Sometimes, fossils in mudrocks become crushed, except for those within concretions that often survive compaction.

Limestones are sedimentary rocks composed mainly of carbonate minerals, especially the calcium carbonate minerals calcite and aragonite, but also sometimes another mineral called dolomite, which is a calcium magnesium carbonate. The main source of carbonate minerals are the shells and other biomineralized skeletons of animals and plants that lived in the sea. These can be intact or fragmented, sometimes into sand- or mud-sized grains. Limestones are a rich source of fossils, some consisting almost entirely of calcareous fossils, as in the Chalk of northern Europe, which is made of the tiny plates of planktonic organisms called coccolithophores (see Chapter 12) that flourished in the sea during Cretaceous times.

Pyroclastics are another kind of sedimentary rock. These consist of fragments ejected by volcanic eruptions, including volcanic tuff. Shells of marine animals can be preserved in tuff produced from ashes ejected by volcanoes. The famous bodies of Pompei citizens, buried by ashes from the eruption of Vesuvius in AD 79, are an archaeological example of preservation in pyroclastic rocks. These chilling remains are displayed in the form of casts made by pouring plaster into voids in the rock where the ash enclosed the bodies.

How likely is it that an organism will become a fossil?

It has been estimated that at least 8 million biological species inhabit the Earth today. Today's biota is the very tip of the biodiversity iceberg – some scientists believe that approximately 99% of species ever to have existed are now extinct. By this reckoning, close to a billion species have at one time or another populated our planet. Only the tiniest proportion of these species are known as fossils.

The incompleteness of the fossil record is not surprising when the multiple requirements for fossilization are considered. For one thing, species lacking a resistant mineralized skeleton are seldom fossilized: some animal phyla (e.g. nematodes or 'round worms') that are very diverse in the present day have no significant fossil record. Furthermore, animals and plants inhabiting some environments are unlikely to become buried in sediment, a normal prerequisite for fossilization. This is true of species from mountainous regions where erosion is dominant and sediment deposition can be minor. Burrowing animals and plant roots entering newly deposited sediments can fragment and destroy any incipient fossils. Shells in marine sediments must pass through a so-called 'taphonomically active zone' (the term 'taphonomy' refers to the processes of fossilization). Here, the decomposition of organic material increases pore water acidity, causing dissolution of the calcareous shells. Burrowing animals may disturb what remains of these shells.

Even after these barriers to fossilization are overcome, there is no certainty that the rock hosting the fossil will survive and not be destroyed by subsequent erosion. The best chance of long-term survival of fossil-bearing rocks is in sedimentary basins where subsidence creates the space that allows thick piles of sediment to accumulate. However, if these rocks are buried too deeply, they may be subjected to intense heat and pressure, obliterating any fossils present. As if these factors conspiring against fossilization were not enough, the probability of a fossil becoming exposed on the surface where it can be found and collected is infinitesimally small. A treasure trove of fossils, many belonging to undescribed species, lies buried in the rocks beneath our feet; most will never see the light of day.

Can ancient communities be fossilized?

Fossils are seldom found singly. More often than not, they occur in 'assemblages' with other fossils belonging to the same or different species. These assemblages allow palaeontologists a glimpse into ancient communities. Some of the best examples come from bedding plane assemblages in which fossils are preserved on the surfaces of beds. Spectacular examples of bedding plane assemblages occur in the Silurian Wenlock Limestone of the English Midlands with brachiopods, bryozoans, corals, trilobites and other shell-bearing marine animals.

But how faithfully do fossil assemblages mirror ancient living communities? It is tempting to interpret them in this way, but there are three main problems. The first is the typical loss of organisms lacking hard parts. For example, Silurian communities from the Wenlock Limestone would originally have included soft-bodied worms of various sorts along with arthropods lacking mineralized skeletons, molluscs with shells of easily dissolved aragonite and probably plants, none of which are fossilized in the bedding plane assemblages. A second caveat is the possibility that the assemblage contains some fossils that were transported to the place of burial by currents, scavengers or predators and were not part of the living community. Finally, it can be difficult to be certain that all the fossils are of organisms that lived at the same time. Is the fossil assemblage a snapshot of a community, or the equivalent of a time lapse containing a mixture of communities, a phenomenon called 'time-averaging'?

ABOVE A Silurian seabed 'graveyard'. This assemblage of fossils includes the shells of brachiopods, as well as branches of coral and bryozoan colonies. From the Wenlock Limestone of Dudley, England, the field of view is 12 cm (4¾ in) across.

How is geological time divided?

Studying the history of human civilizations would be impossible if we did not know the order in which major events happened, or when key people lived. Likewise, if we are to use fossils to reconstruct the history of life on our planet, we need to place them in their correct time context. Just as British historians think in terms of time periods, such as Tudor, Elizabethan, Georgian and Victorian, so palaeontologists employ geological periods as principal markers of time. Although the names of most of these geological periods may be unfamiliar – with the probable exception of the Jurassic, thanks to the film *Jurassic Park* and its spin-offs – they are crucial when discussing fossils.

The geological timescale was assembled piecemeal, predominantly by early nineteenth century geologists studying sequences of rocks and their constituent fossils in different parts of Europe. Many of the names of the geological periods that these geologists proposed were derived from the geographical regions where they worked. For instance, the Cambrian period was founded by Adam Sedgwick who was studying rocks in Wales (named *Cambria* in Latin), the Permian period was proposed by Roderick Murchison who researched the geology around Perm in the Ural Mountains of Russia, and the name Jurassic was coined by Alexandre Brongniart from the Jura Mountains on the border of France and Switzerland. A few new geological periods have since been added by stratigraphical committees to categorize some older rocks, including the Ediacaran for the period when the first abundant animal fossils appeared. Although the sequence of geological periods has been known for 200 years, the absolute dates of the beginnings and ends of the periods did not become known until the development of radiometric dating, as explained below. Even now, these dates are subject to minor modifications as better radiometric age estimates become available. In addition, because geological periods were established with no notion of absolute time, they vary in their durations, from approximately 25 to 80 million years.

Geological periods are grouped into larger units of time called eras and eons, and are subdivided into series (or epochs) and stages. The great majority of fossils occur in rocks of the Phanerozoic eon, which encompasses the Cambrian and all younger geological periods and covers the last 539 million years of geological time.

Three eras – Palaeozoic, Mesozoic and Cenozoic – are recognized in the Phanerozoic. Like geological periods, these eras were first named around 200 years ago, based on major differences in the fossils they contain. Their names mean old, middle and new life, respectively. The differences between the fossils characterizing the three eras are in large part due to mass extinctions at the ends of the Palaeozoic and Mesozoic eras. For instance, trilobites are typical of the Palaeozoic but became extinct during the end-Permian mass extinction marking the close of the era, whereas ammonites characterize the Mesozoic before their demise during the Cretaceous mass extinction at the end of that era. Preceding the Phanerozoic is a long interval of time informally known as the Precambrian and consisting of three eons: Hadean, Archean and Proterozoic.

Finer divisions of geological time are needed to place rocks and fossils into a more precise timeframe, which is where things get a little more complicated. Some geological periods, such as the Jurassic, are divided into three series: Early, Middle and Late. Others, such as the Cretaceous, are just divided into two: Early and Late. The three youngest periods – Palaeogene, Neogene and Quaternary – are subdivided into seven epochs: Paleocene, Eocene, Oligocene, Miocene, Pliocene, Pleistocene and Holocene.

EON	ERA	PERIOD OR EPOCH	AGE (millions of years)
PHANEROZOIC	CENOZOIC	Holocene	0.012
		Pleistocene	2.6
		Pliocene	5.3
		Miocene	23
		Oligocene	34
		Eocene	56
		Paleocene	66
	MESOZOIC	Cretaceous	145
		Jurassic	201
		Triassic	252
	PALAEOZOIC	Permian	299
		Carboniferous	359
		Devonian	419
		Silurian	444
		Ordovician	485
		Cambrian	539
PRECAMBRIAN		Ediacaran	635
		Cryogenian	720
			4600

LEFT Geological timescale, with the dates of the boundaries between the various periods or, in the case of the Cenozoic, epochs.

1. The vertical (time) axis is not to scale
2. Only the two youngest periods and none of the eras of the Precambrian are shown
3. Epochs rather than periods are specified for the Cenozoic

Periods and epochs are further subdivided into stratigraphical stages. For instance, the Jurassic period comprises 11 stages, each lasting on average 5 million years. These generally take their names from places in which the rocks of the stage were first studied, e.g. Bathonian from Bath, and Oxfordian from Oxford. One of the goals of stratigraphers who study the succession and ages of rocks has been to identify geological reference sections where the formal bases of these stages are best placed, so-called 'golden spikes'. If you find all this a bit challenging, then you are not alone. Even experienced palaeontologists need to consult the charts, which are updated annually by the International Commission on Stratigraphy (www.stratigraphy.org) in which all divisions are given along with the times when they started and ended.

Stratigraphical nomenclature makes a distinction between units of geological time and the rocks and fossils from these intervals of time. For instance, the Jurassic System comprises the rocks formed during the Jurassic Period, as well as the fossil remains in these rocks of organisms that lived at the time. To draw a crude analogy from human history, the Victorian period in British history is defined by the reign of Queen Victoria (1837–1901), which is distinct from the products of that time, such as Victorian coins, buildings and furniture.

A hierarchical system is used to name stratified rocks: a stratigraphical group contains two or more formations, and a formation contains two or more members. For example, rocks in the English Midlands deposited during latest Triassic and earliest Jurassic time belong to the Lias Group, the oldest subdivision of which is called the Blue Lias Formation and contains three members: from oldest to youngest, these are the Wilmcote Limestone Member, Saltford Shale Member and Rugby Limestone Member. The member names in this case are useful in showing that deposition changed from limestone to shale and back to limestone. Formations are the most often cited rock units. Just as biological species are defined by a type specimen, so formations and members have a type locality serving as a standard reference point. To place a fossil into context, it is helpful to include not only its geological age but also the formation from which it came. This is particularly true when new research means that the age of the formation changes.

How do we know the age of rocks and fossils?

Unlike coins or hallmarked sterling silver, rocks and fossils do not come stamped with a date. Instead, their ages must be determined indirectly by radiometric dating. Developed by the New Zealand physicist Ernest Rutherford at the beginning of the twentieth century, radiometric dating hinges on certain elements containing unstable, radioactive isotopes that decay at constant and known rates. Take the element carbon. Atoms of the commonest isotope of carbon, carbon-12 (^{12}C) which is stable, contain six protons and six neutrons. Another isotope (^{14}C) also has six protons but there are eight neutrons, making it unstable. This isotope decays into nitrogen (^{14}N) at a constant rate measured by its half-life – the time taken for half of the ^{14}C atoms to decay – shown experimentally to be 5730 years. Relative to ^{12}C, the less ^{14}C there is in a material, such as a bone, tooth, shell or piece of wood, the older it is.

The relatively short half-life of ^{14}C means that this method, often referred to as radiocarbon dating, has limited geological use because there is too little ^{14}C to be detected in material more than 50,000 years old. However, the radioactive isotopes of some other elements have much longer half-lives, allowing them to be applied in the dating of even the oldest rocks. For example, the potassium–argon and uranium–

lead dating methods employ isotopes with half-lives of 700 million years and 1,300 million years, respectively. These methods generally involve dating inorganic minerals such as glauconite or zircon rather than the fossils themselves. Glauconite is a clay mineral that grows in slowly deposited sediments, which often also contain fossils. In contrast, zircon is a mineral found in igneous rocks, including lavas, and volcanic ashes generally devoid of fossils. To age fossils using radiometric dates obtained from zircons requires a process of 'bracketing'. For example, imagine a layer of sedimentary rock containing fossils sandwiched between two ash deposits, the underlying ash radiometrically dated as 32 million years old and the overlying ash as 30 million years old. Bracketing would allow the sedimentary rock, and its constituent fossils, to be estimated as between 30 and 32 million years old.

Unfortunately, not many fossil-bearing rock sequences contain either directly datable glauconite or interbedded volcanic rocks with zircons. In such cases, these sequences have to be correlated with others formed at the same time and whose age is already known. Correlation is traditionally achieved using fossils, although geochemical and palaeomagnetic methods are also increasingly used.

How are fossils used to correlate rocks of the same age?

The first published geological map covering an entire country was created by the English canal and drainage engineer William Smith (1769–1839). Published in 1815, Smith's geological map of England, Wales and parts of Scotland depicts the geographical distributions of different layers of rock (strata) using different colours and bears a close resemblance to modern geological maps of Britain. Smith managed to draw his map because he could match (correlate) strata from one place to another, thanks to his observation that different associations of fossils characterized different strata. Smith established the important principle of recognizing strata by the fossils they contain. However, he worked in the days before evolution became accepted, and when the Earth was thought to be much younger than it is, believing that each stratum represented a separate creation of life that was extinguished at the end of deposition of the stratum, following an idea promoted by the great French naturalist Baron Georges Cuvier (1769–1832).

During the 200 years after Smith, geologists have shown how fossils can be used to correlate strata across the entire world using fossils. Some fossils are much better than others in this respect and are used as guide fossils to define stratigraphical zones. Ideally, guide fossils are widely distributed geographically and evolve rapidly, so each species has a short duration. This then allows correlation at a fine timescale. For sedimentary rocks deposited in marine environments, fossils of planktonic organisms and free-swimming oceanic animals tend to have wider geographical ranges than those living immobile on the seafloor and therefore make better guide fossils. Ammonites were free-swimmers and fulfil the requirement of evolving rapidly (see Chapter 4), so they are often used as guide fossils for correlation. For example, 145 ammonite zones and subzones have been identified in the Jurassic of the UK, each averaging less than half a million years in duration. Microfossils, such as foraminifera (see Chapter 12), are also used frequently in correlation because they have the advantage of being exceedingly abundant – the chance of finding a tiny microfossil in a drill core is far greater than the core fortuitously intersecting a macrofossil such as an ammonite.

Imagine that a hypothetical sedimentary rock, radiometrically dated at between 30 and 32 million years old based on associated volcanic ashes, contained fossils potentially useful for correlation. This would allow rocks elsewhere containing the same fossils also to be dated as 30–32 million years old. Through the efforts of countless geologists around the world radiometrically dating rocks and using fossils and other means to correlate strata from one place to another, a vast amount of information has been generated. This has led to refinements in the geological timescale and a more complete understanding of the ages of fossils, and consequently the history of life.

Global changes and life of the past

If the fossil record tells us anything it is that the inhabitants of our planet have changed enormously through geological time: life has been in a state of constant flux with species appearing as others have become extinct. This evolutionary story has unfolded against an unremitting backdrop of global environmental changes, with climates becoming warmer or colder, sea levels rising or falling, areas of land emerging as the sea retreats before being submerged again, mountain ranges rising and then being slowly eroded into plains, and continents wandering across the globe from poles to equator. Although these changes have typically taken place slowly over geological time, major volcanic eruptions and impacts by asteroids have sometimes brought about rapid changes with profound effects on life, not just locally but on a global scale.

The revolutionary theory of plate tectonics sent reverberations through the geological sciences in the 1960s. Plate tectonics describes the Earth's crust and upper mantle – together forming the lithosphere – as a series of plates that move slowly over the underlying asthenosphere. Plate tectonics succeeded in explaining numerous phenomena, such as the apparent fit between the continents on each side of the south Atlantic Ocean, suggesting that they had at one time been joined; the peculiar geographical distributions of some living and fossil organisms that made little sense based on present-day geography; the symmetrical stripes of igneous rocks showing

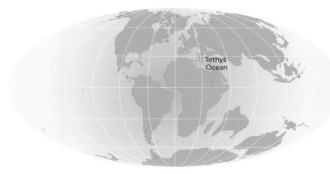

LEFT Palaeogeographical map of the world 95 million years ago during the mid Cretaceous. The Atlantic Ocean was beginning to open but the Indian subcontinent was still in the southern hemisphere at this time.

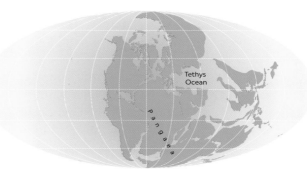

RIGHT Palaeogeographical map of the world 248 million years ago during the Early Triassic at the dawn of the Mesozoic Era. Most of the landmasses were aggregated into the supercontinent Pangaea.

LEFT Palaeogeographical map of the world 340 million years ago during the Carboniferous period when low-lying landmasses were flooded.

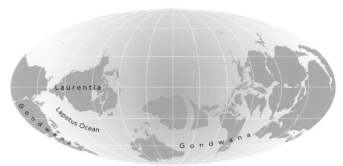

RIGHT Palaeogeographical map of the world 525 million years ago during the Cambrian period. Most of the continents at this time were widely dispersed across the equator and southern hemisphere.

normal and reverse magnetization on either side of mid-oceanic ridges, and the linear patterns of volcanoes and earthquake zones.

With a maximum value of about 10 cm (4 in) per year, the rate of plate movement may seem trivial, but over the enormity of geological time it has brought about huge changes in the positions of the continents. Numerous lines of evidence have been used by geologists to construct palaeogeographical maps showing how the continents have drifted across the surface of the globe over hundreds of millions of years. Land masses have become separated by newly formed oceans, dividing populations of land-dwelling animals and plants, and have been reunited when these oceans closed, bringing together terrestrial biotas that had evolved independently. At the same time, new oceans have linked previous oceans, changing patterns of current flow and allowing marine organisms to become more widely dispersed, while closure of oceans has separated marine populations.

The location of the continents over the globe has varied enormously through geological time. Most of the continents have been positioned in the mid to high latitudes of the southern or northern hemispheres during some times, but at other times they have straddled the tropics. These differing configurations have had a major effect on climates. When continental masses are positioned over the poles, there is the potential for thick masses of ice to develop from the snow falling on them, locking up water and leading to a drop in global sea level.

Processes occurring at plate boundaries have profound effects on global topography, and consequently on the range of environments available for life to colonize. Mid-oceanic ridges mark divergent plate boundaries where new oceanic crust is formed. These ridges become bigger with increasing rates of seafloor spreading, displacing water and causing sea levels to rise. A rise in sea level floods the lower-lying areas of the continents with shallow seas that are inhabited by many of the animals commonly found as fossils. Places of maximum sediment deposition – known as sedimentary basins – are often located where plate movements have pulled the crust apart, resulting in the subsidence needed to provide the space where sediments containing fossils can accumulate. Convergent plate boundaries often involve an oceanic plate sinking beneath a continental plate, a process known as subduction. An island arc of volcanoes develops here, and the sediments scraped from the subducted plate accumulate as an 'accretionary prism' of rocks. Crustal thickening and mountain building occur at these locations, with sedimentary rocks containing the fossils of animals that lived on the seafloor being thrust upwards to form high mountains.

Over the last 600 million years, Earth's climate has alternated several times between warm and cold intervals, known as greenhouse and icehouse worlds, respectively. Today, we live in an icehouse world. However, greenhouse worlds have predominated

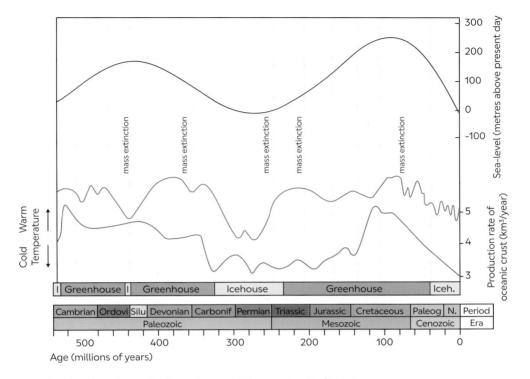

ABOVE Our planet has witnessed major environmental changes since the Cambrian period. These include rises and falls in average sea-level (upper curve), rate of production of new oceanic crust (middle curve), and global temperature (lower curve) which has entailed alternations between warm 'greenhouse' and cold 'icehouse' worlds. There have been five large mass extinction events during this same interval of time.

through geological time. During these times there are no polar ice-caps, the mean global temperature hovers around 21–22°C (70–72°F) (compared with 14–15°C (57–59°F) today) and the distinction between seasons is reduced. The last time the Earth was in a greenhouse state was at the end of the Eocene epoch when the high temperatures allowed palm trees to grow in the Arctic and conifer forests to spread across parts of Antarctica. Because water was not locked up in ice-caps and glaciers, global sea levels were hundreds of metres higher in greenhouse worlds. High sea levels result in a smaller habitable area of land surface, but they also produce vast expanses of seas over the edges of the continents in which shallow-water marine animals thrive, and these are the mainstay of the fossil record. Greenhouse conditions are believed to have existed when volcanic activity was intense, which is generally coincident with high rates of seafloor spreading, releasing extra amounts of the greenhouse gas carbon dioxide into the atmosphere.

Not all globally important geological processes have occurred at plate boundaries. Occasional eruptions of huge amounts of basaltic lavas are believed to have resulted from mantle plumes creating hotspots beneath crustal plates. Vents punched through the crust allowed highly mobile 'flood basalts' to spread hundreds of kilometres over the land surface, accumulating over time into piles of basalt lava flows that can be more than 1,000 m (3,280 ft) thick. These are often known as 'traps' from the Swedish word *trappa* or staircase, referencing the step-like topography they produce. The effects of such enormous outpourings of lava include the release of gases such as carbon dioxide and sulphur dioxide into the atmosphere, with important consequences for living organisms. Indeed, some mass extinction events have been linked to flood basalts: the Siberian Traps were formed at more-or-less the same time as the end-Permian mass extinction, and the Deccan Traps of India coincided with the end-Cretaceous (K–Pg) mass extinction.

An alternative cause of the K–Pg mass extinction event, which extinguished the non-avian dinosaurs, ammonites and many other animals and plants, was suggested in a ground-breaking paper published in 1980 by a team led by Luis Alvarez and his son Walter. This paper reported high values of the element iridium in thin layers of sedimentary rocks coincident with the extinction level in geological sections as far apart as Denmark, Italy and New Zealand. As iridium is rare in terrestrial rocks but can be abundant in meteorites, it was inferred that a large extraterrestrial object (bolide) had struck the Earth, triggering a cascade of environmental catastrophes, which ultimately led to the rapid extinction of many species that had prospered before

RIGHT Stevns Klint in Denmark exposes rocks spanning the K–Pg boundary, marking one of the major mass extinctions in the history of life. The geological hammer rests on the top of the Cretaceous Chalk above which there is a thin layer of clay, rich in iridium, derived from the impacting bolide.

LEFT *Hoploscaphites* is one of the last ammonites before the K–Pg mass extinction. Indeed, it is possible that some populations dwindled on for a few thousand years into the Paleocene. This specimen is from the Late Cretaceous Fox Hills Formation of Dewey County, South Dakota, USA.

the impact. In a story now well-known, subsequent research pinpointed the site of bolide impact 66 million years ago as Chicxulub on the Yucatan Peninsula of Mexico. Most palaeontologists now accept this impact as instrumental in the K-Pg extinction, although some believe that the 'double whammy' of an impact against a background of the global stress caused by the gasses emitted from the Deccan Trap volcanoes was needed for such devastation of the Earth's biota.

In addition to the K–Pg extinction, four other major mass extinctions have been recognized since the Cambrian period, at or near the ends of the Ordovician, Devonian, Permian and Triassic periods. The Ordovician mass extinction was coincident with pulses of global cooling and glaciation. Two separate events contributed to the Late Devonian mass extinction, but their causes are still being debated. The 'mother of all mass extinctions' obliterated an estimated perhaps 90% of living species at the end of the Permian period, including the last remaining trilobites and blastoid echinoderms. Some scientists believe that life on Earth had a close shave with total annihilation. Feasibly driven by the massive volcanic eruptions that created the Siberian Traps, this was a time of extreme and prolonged global warming when the land surface became arid, and the oceans became hot, stagnant and acidic. The cause of the end-Triassic mass extinction remains uncertain, although many geologists believe it was related to enhanced vulcanicity associated with the opening of the Atlantic Ocean.

2 The oldest fossils

The Earth, according to current estimates, is 4,543 million years old. Initially inhospitable to life, within a mere 1,000 million years or so living organisms had populated our planet. Evidence for animals and plants of a macroscopic size is first found in rocks some 2,000 million years later, just before the beginning of the Cambrian period. The last 540 million years have witnessed an evolutionary rollercoaster with the appearance of an increasing variety of larger and more complex life forms.

When Charles Darwin published *On the Origin of Species* in 1859 fossils were almost unknown from the enormous interval of time called the Precambrian, extending from the Earth's origin to the beginning of the Cambrian period. The absence of Precambrian fossils was described by Darwin as 'inexplicable'. He even saw it as an obstacle to the acceptance of his theory of evolution. Darwin was right to be cautious. Since his death, however, numerous fossils have been discovered in Precambrian rocks, allowing an intriguing glimpse into life during Earth's early history. Hugely important evolutionary innovations occurred during the Precambrian, with the emergence of many things that we take for granted in life today. These include organisms with nuclei (eukaryotes) and with more than one cell (multicellularity), as well as photosynthesis, predation and sex. Fossils may not help much in solving the vexed question of the actual origin of life, but they do tell us a lot about life's early evolution and the inter-relationships between life and conditions on the Earth's surface.

The Precambrian is subdivided into three eons: Hadean (4,567–4,000 million years ago), Archean (4,000–2,500 million years ago) and Proterozoic (2,500–540 million years ago). There is no evidence for the existence of life during the Hadean eon, although it must be noted that rocks of this age are exceedingly rare. The oldest fossils and chemical indicators of life have been discovered in rocks from the Archean eon. At this point it is worthwhile emphasizing the difficulties entailed in finding, recognizing and interpreting Archean fossils, which are a far cry from

OPPOSITE Reconstruction of animals living close to the seafloor during the deposition of the Middle Cambrian Burgess Shale of British Columbia, Canada. The large animal is a radiodontid arthropod, which towers above two spiny individuals of *Hallucigenia* perambulating over the seabed.

the larger, more complex and unambiguous fossils found in Phanerozoic rocks. For one thing, most Archean rocks have had plenty of time to be cooked and crushed. Such metamorphism destroys, or at best degrades, any fossils these rocks might have contained. Second, early life forms did not generally make hard, mineral skeletons, lessening their chances of being fossilized. Third, all Archean and most Proterozoic fossils are microfossils, which measure less than a millimetre in size and can be easily overlooked. The study of Precambrian microfossils normally requires processing of rocks for observation by microscopy; either the preparation of wafer-thin sections of rock through which light will pass, or the use of chemicals such as hydrofluoric acid to dissolve the rock and separate out the tiny fossils contained within. Finally, nearly all Precambrian microfossils have simple shapes: spheres (coccoids), rods, filaments, etc. These can be difficult to distinguish from structures created by inorganic minerals. Additional evidence, such as chemical composition, may be required to discriminate between biological (biogenic) microfossils and non-biological (abiogenic) structures.

Ancient microbial fossils

It may seem incredible that tiny bacteria and other microbes are ever fossilized at all. However, several forms of early mineralization can allow their preservation. These include: (1) silicification, in which silica is precipitated onto the cell walls before they decay; (2) phosphatization, in which the tissues are replaced by tiny crystals of apatite; and (3) calcification, entailing calcium carbonate mineralization, often by the microbes themselves, within or on the surfaces of their cells. Silicified microbes are typically found in the siliceous rock chert, whereas calcified microbes often occur in layered structures called stromatolites, which are described below (p. 32). Remnants of the organic walls of microbes also sometimes survive as compressed carbonaceous fossils, contrasting with the three-dimensionally preserved microfossils preserved through silicification.

What are the oldest fossils ever discovered? In 1993 Bill Schopf, an internationally renowned authority on Precambrian life, described filamentous structures in the Apex Chert of Western Australia dated as 3,465 million years old. Measuring just 0.5–19.5 microns in diameter, these filaments were interpreted as fossilized microbes by Schopf, potentially making them the oldest fossils known. Not all palaeontologists have accepted Schopf's interpretation, with some considering the Apex Chert filaments to be hairline fissures in the rock that later became infilled by minerals. A biological origin is supported by dark coatings of apparent kerogen, a form of organic carbon, around the filaments. However, it is possible that this is carbon that seeped into cracks in the rocks. For now, the biological origin of the Apex Chert filaments

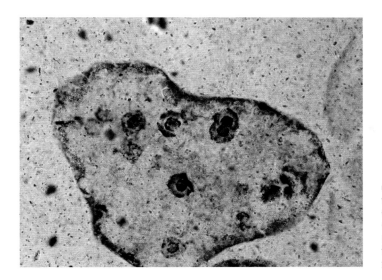

LEFT Spherical microbial fossils visible in a thin section of Gunflint chert from Ontario, Canada. Measuring just 25 microns in diameter, these fossils are almost 1,900 million years old.

remains contentious. Nevertheless, structures more widely accepted as microbial fossils have been described from deposits only slightly younger in age. These include carbonaceous spheres in the Strelley Pool Formation, also from Western Australia, which is roughly 3,430 million years old.

All Archean fossils are prokaryotes, that is single-celled organisms lacking a nucleus and other membrane-bound organelles, the subunits within cells that have specific functions. Most of these fossils appear to have been bacteria, but some probably belonged to another major group of prokaryotes confusingly called 'Archaea'. These microbes employed a wide variety of metabolic processes to generate the energy to live. Importantly, a group of bacteria now called cyanobacteria, but formerly known as blue-green algae, were photosynthetic organisms, which generated oxygen as a by-product of this process. Although the oldest unambiguous fossil cyanobacteria are about 1,900 million years old, it is likely that they had a considerably longer evolutionary history.

Geochemical evidence shows oxygen to have been extremely sparse in the atmosphere and oceans during the Archean, with atmospheric oxygen a mere 0.001% of today's levels. Although the photosynthetic activities of cyanobacteria introduced oxygen into the atmosphere, most of this was initially soaked up by reactions with volcanic gases in the atmosphere and iron in the oceans. Banded ironstone formations formed when dissolved iron was oxidized to produce particles of insoluble iron oxide minerals that sank to the seabed. By 2,450 million years ago, the soluble iron needed for these distinctive and economically important deposits had been largely exhausted and oxygen levels in the sea and atmosphere began to rise significantly in what is known as the Great Oxidation Event, which lasted until about 2,100 million years ago. This had enormous consequences for Earth's biosphere. In particular, it paved the way for the evolution of many kinds of eukaryotic organisms with nuclei and membrane-bound organelles requiring large amounts of oxygen for respiration (see below, p. 34).

Stromatolites

Although life was copious on the young planet Earth, until late in the Precambrian all individual organisms would have been too small to see with the human eye. However, their presence would have been manifest in reef-like structures called stromatolites constructed by communities of micro-organisms. Stromatolites are typically internally layered or laminated and usually made of calcium carbonate, although in some cases this has been replaced by silica to form chert stromatolites. Most stromatolites are mound shaped, with individual mounds often coalescing at their edges, but others are flat or have the shape of columns or tall pinnacles. There are several places in the world today where living, growing stromatolites can be seen. The most famous of these is Hamelin Pool in Shark Bay, Western Australia, a

BELOW Vertically sectioned stromatolite from Siberian rocks dated as more than 2,000 million years old. Each column is about 1.5 cm (½ in) wide on average.

ABOVE Modern columnar stromatolites growing in the shallow waters of Hamelin Pool, Shark Bay, Western Australia.

World Heritage Site and tourist attraction. The stromatolites here are accompanied by other microbial-formed structures called thrombolites which lack the layered internal structure of stromatolites. The stromatolites in the shallowest waters of Hamelin Pool are flat sheets constructed by mats of filamentous cyanobacteria, whereas those from deeper waters are columnar and formed by mats of coccoid cyanobacteria. Both types of cyanobacteria precipitate fine grains of calcium carbonate which quickly become cemented together into solid structures. The cyanobacteria and other microbes also contribute to stromatolite formation by trapping and binding particles of sediment from the water. At rates of less than 0.5 mm per year, upward growth of the stromatolites is very slow. Stromatolites have been present in Hamelin Pool for the last 2,000 years and some individual columns have been growing for a thousand years.

Stromatolites have a very long geological history – the oldest known structures interpreted as stromatolites come from the Strelley Pool Formation of Western Australia, which is dated at approximately 3,430 million years old, as mentioned above. Unfortunately, microbial fossils have not been observed in the stromatolites of the Strelley Pool Formation and are scarce in other examples of Archean stromatolites, leaving a question mark over the biological origin of these laminated structures. The same

ABOVE The polished limestone used for the columns of the Great Hall of the People in Beijing, China is full of Proterozoic stromatolites, each a few centimetres in width and with alternating layers red and white in colour.

doubt does not apply to many Proterozoic stromatolites, which can be full of microbial fossils, as in the Gunflint chert of Canada, a 1,880 million-year-old rock containing silicified fossils of filamentous and coccoidal microbes that are most likely cyanobacteria. The heyday of stromatolites occurred in the mid Proterozoic before a decline in the late Proterozoic that is often attributed to destructive grazing by newly evolved animals, although the rise of seaweeds might also have played a role.

There is no need to venture into remote parts of Australia or Canada to see Precambrian stromatolites. Visitors to Tiananmen Square in Beijing can marvel at the stromatolites in the carved red limestone decorating the bases of the massive columns of the Great Hall of the People. These stromatolites come from the Northeast Red Formation of eastern China, a deposit that is about 850 million years old.

Eukaryotes, multicellularity and sex

Eukaryotes are characterized by membrane-bound organelles and have cells of larger size than prokaryotes. All animals, plants and fungi are eukaryotes. It is now generally accepted that eukaryotes evolved through symbiosis between different types of prokaryotes. Particularly associated with the brilliant evolutionary biologist Lynn Margulis (1938–2011), this theory of 'symbiogenesis' proposes that the organelles in eukaryote cells were derived from various kinds of bacteria that became incorporated into host cells. For example, the chloroplasts used in photosynthesis came from cyanobacteria, while the mitochondria used in respiration were derived from aerobic bacteria. The presence in mitochondria of DNA similar to that of some bacteria is a legacy of their origin from these prokaryotes.

Molecular clock estimates (p. 43) for the origin of eukaryotes differ enormously but the oldest eukaryote fossils are about 1,650 million years old. Many early eukaryotes comprise the resting cysts of plankton and are covered with spines or otherwise ornamented. These are usually placed into the taxonomic wastebasket group called acritarchs (p. 235). Other early eukaryotes were benthic and lived on or close to the seafloor. Eukaryotes were once thought to be rare in rocks more than about 800 million years old. However, new fossil discoveries show that eukaryotes became quite diverse not long after they are first recorded as fossils.

A particularly important Proterozoic eukaryote is *Bangiomorpha* from rocks in Arctic Canada that are slightly over 1,000 million years old. *Bangiomorpha* has filaments about 25 microns in width consisting of multiple cells and was attached to the seafloor using a well-defined root-like holdfast. The presence of distinct reproductive cells makes *Bangiomorpha* the oldest fossil showing evidence of sexual reproduction, as well as one of the oldest with multicellularity. Similarities between *Bangiomorpha* and the living red alga *Bangia*, from which its name was derived, suggest that *Bangiomorpha* was also a red alga or rhodophyte.

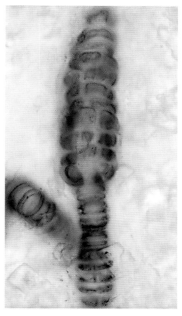

ABOVE Visible in a thin section cut from a rock over 1,000 million years old, *Bangiomorpha* is believed to be a red alga with filaments 30 microns in diameter. Somerset Island, Arctic Canada.

Ediacaran fossils

In 1957 Roger Mason made an unexpected discovery when rock climbing in Charnwood Forest, Leicestershire, UK along with some schoolboy friends. He found an enigmatic frond-like fossil with a patterned surface in the late Precambrian rocks exposed there. This fossil was subsequently named *Charnia masoni* in his honour. Eleven years earlier, prospecting geologist Reg Sprigg was equally surprised to find fossils in rocks of similar age in the Ediacara Hills of South Australia. Sprigg later interpreted these fossils as jellyfishes, proposing several new genera including *Dickinsonia*, described below. Both the Leicestershire and South Australian fossils are preserved as imprints in fine-grained sandstones and lack biomineralized skeletons. Late Precambrian macrofossils like these were dubbed vendobionts or Ediacarans. They characterize a geological period formally recognized in 2004 called the Ediacaran and lasting from 635 to 540 million years ago. Ediacaran fossils have been collected from other parts of the world too, notably Newfoundland in Canada, Namibia, and the region of the White Sea in Russia.

Perhaps the most emblematic fossil from the Ediacaran is *Dickinsonia*, which even featured on a 50-cent postage stamp issued by Australia in 2005. Vaguely resembling a flatworm but with an elliptical body divided into narrow, segment-like units, fossils of *Dickinsonia* exhibit a remarkable range in size, from less than 0.5 cm to more than 80 cm long (from ⅕ in to more than 31½ in). The units radiate from a medial furrow or ridge that extends most of the way along the length of the body. During growth, new units were apparently added at one end to extend the length of the body while existing units grew to broaden the body. The largest examples contain 375 body units. Trails left on bedding planes show *Dickinsonia* to have been mobile, implying that it was an animal rather than a plant. Its animal nature has been corroborated by the discovery of cholesterol-like chemicals associated with specimens of *Dickinsonia*, which are biomarkers for animals. *Dickinsonia* apparently lacked a gut and instead may have employed external digestion, possibly feeding on dissolved organic carbon. Like so many Ediacaran fossils, it has been impossible to place *Dickinsonia* into a modern animal phylum.

While *Dickinsonia* had the ability to move, many other Ediacarans were static, including *Charnia* and similar frond-like fossils grouped together as rangeomorphs. Although the great majority of rangeomorphs are of Ediacaran age, at least one genus (*Stromatoveris*) is known from the Cambrian, showing that these peculiar creatures survived beyond Ediacaran times. Like *Dickinsonia*, it has been proposed

ABOVE Individuals of two species of the enigmatic Ediacaran fossil *Dickinsonia* from the Rawnsley Quartzite of South Australia. The smaller fossil is *Dickinsonia tenuis* 3 cm (1 in) across, the larger is *Dickinsonia costata* 4.5 cm (1¾ in) across.

that rangeomorphs were also animals that fed on dissolved organic carbon. Some, such as *Fractofusus*, exhibit fractal branching patterns with several orders of branches of ever-decreasing size ensuring a large surface area for feeding. *Fractofusus* is found in Newfoundland where the fossil-bearing Ediacaran rocks were deposited in a deeper water environment than those from South Australia.

The preservation of so many soft-bodied fossils in the Ediacaran seems puzzling but may be related to the common presence of microbial mats covering the seafloor at this time. Microbial mats are multilayered sheets of bacteria, including cyanobacteria, and other micro-organisms. The stromatolites described above are a kind of cemented microbial mat. It has been proposed that the preservation of soft-bodied animals such as *Dickinsonia* resulted from the periodic deposition of sand, or occasionally volcanic ash, smothering the microbial mats and the larger animals living on them. This was often followed by rapid bacterial precipitation of iron minerals to form what have been dubbed 'death masks' of the Ediacaran animals. The Ediacaran microbial mats were eaten by some of the mobile animals possessing guts (e.g. *Kimberella*, believed to be a primitive mollusc), while 'mining' of the mats is evident from the occurrence of the burrowing trace fossil *Helminthoidichnites*. Microbial mats of the kind found in the Ediacaran subsequently became much less common following the proliferation in the Cambrian of grazing animals capable of destroying them.

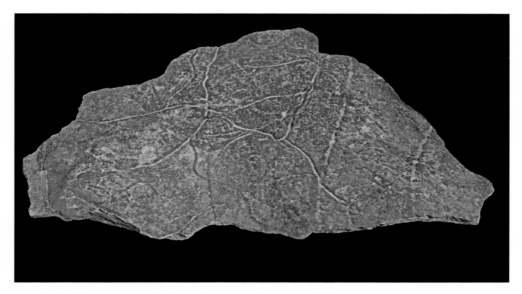

ABOVE Angled view of a meandering horizontal burrow system assigned to the trace fossil *Helminthoidichnites* from Ediacaran rocks of Bathtub Gorge, South Australia. Rock 20 cm (7¾ in) across.

RIGHT Field photograph of limestone from the Ediacaran of Namibia packed with calcareous tubes of *Cloudina*. Each tube is about 5 mm in diameter in this early example of an animal with a biomineralized skeleton.

In contrast to most Ediacaran fossils, which are soft-bodied, *Cloudina* has a skeleton of calcium carbonate and is among the earliest known animals possessing a biomineralized skeleton. Named in honour of Preston Cloud (1912–1991), a pioneer in Precambrian palaeontology, *Cloudina* consists of tubes a few millimetres in diameter, up to 15 cm (6 in) long, often sinuous, occasionally branching, and typically occurring along with others in clusters. Transverse flanges on the exterior of the tubes reflect the diagnostic internal construction of *Cloudina*, which comprises a stacked series of elongate cones, each cone representing a new increment of growth. The phylogenetic relationships of *Cloudina* have yet to be settled, with some scientists favouring an annelid affinity while others believe it to have been a coral-like cnidarian. Regardless, *Cloudina* is likely to have been a sessile suspension feeder capturing plankton by employing tentacles protruding from the open end of the tube. In the Nama Group of Namibia, *Cloudina* is found in limestones deposited in deeper waters offshore from the sandstones containing soft-bodied fossils like those mentioned above.

Finally, astonishing embryo-like fossils of Ediacaran age, first reported in 1998, are abundant in the Doushantuo Formation of China. Preserved as phosphatic replacements of soft tissues, the Doushantuo fossils are balls consisting of a few to several thousand cells and seemingly show different stages in embryo cleavage from one to two to four to eight cells, and so on. However, it is unclear in the absence of adults exactly what organisms they represent. Some believe them to be animal embryos, others the embryos of plants, fungi or even a kind of bacteria. Should

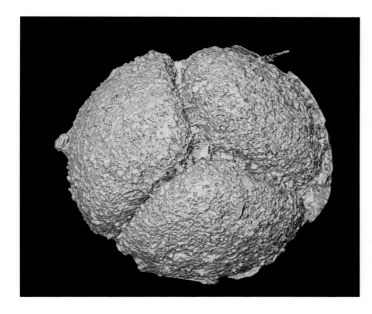

LEFT The Ediacaran fossil *Tianzhushania* represented by a cluster of cells from the Ediacaran Doushantou Formation of Guizhou Province, China, and imaged using an X-ray tomographic microscope. About 700 microns in diameter.

further research on these extraordinary fossils show them to be unequivocal animal embryos, they can lay claim to being among the oldest known animal fossils because the Doushantuo Formation is of early–mid Ediacaran age and directly overlies glacial deposits of the Cryogenian period making it older than the Ediacaran fossils such as *Charnia* and *Dickinsonia* described above.

The 'Cambrian Explosion'

Anyone who has looked for fossils in Cambrian rocks will know how difficult it is to find them. Although trilobites (see Chapter 4) and shell-bearing animals such as brachiopods (see Chapter 5) can be found at some localities, Cambrian fossils are not nearly as abundant or diverse as are fossils in Ordovician and younger rocks. Nevertheless, the oldest known examples of all routinely fossilized animal phyla occur in Cambrian rocks. The sudden appearance of fossils belonging to so many phyla has been termed the 'Cambrian Explosion'. Before considering the meaning of the Cambrian Explosion, it is necessary to look at some iconic and important Cambrian fossils.

Animals with mineralized skeletons were not the only ones to make their debuts in the Cambrian fossil record. Contrary to his middle name, Charles Doolittle Walcott (1850–1927) was an immensely productive and innovative palaeontologist who rose to become the Secretary of the Smithsonian Institution. In 1909, Walcott discovered the Burgess Shale, a Middle Cambrian Lagerstätte, in British Columbia, Canada.

Remarkably diverse soft-bodied fossils were found in the Burgess Shale by Walcott, and many additional species have been collected since, brought to the attention of the public by Stephen J. Gould (1941–2002) in his book *Wonderful Life*. But the Burgess Shale is not the only Cambrian fossil treasure house. Now a UNESCO World heritage site, the Cambrian Lagerstätte at Chengjiang in Yunnan, China was discovered in 1984 by Hou Xian-guang. Almost 200 species have been found at Chengjiang which, dated at 520 million years ago, is 12 million years older than the Burgess Shale.

Fossils from these two and other Cambrian Lagerstätten, such as those in Emu Bay, South Australia and Sirius Passet, Greenland, provide invaluable windows on the communities populating the oceans of the Cambrian period. Compared with the enigmatic fossils from the Ediacaran period described above, these Cambrian animals much more closely resemble animals living today in morphology. The majority can be readily placed into a modern phylum, although there is no shortage of Cambrian oddities still without a home. Arthropods are the most diverse phylum in the Burgess Shale and at Chengjiang, but a range of animal phyla are represented, including sponges, ctenophores, priapulid and annelid worms, brachiopods, chaetognaths, hemichordates and chordates. Other Cambrian fossils belong within the 'stem-groups' of extant phyla. Stem-group fossils possess some but not all of the attributes

BELOW *Hallucigenia* is a Cambrian lobopodian, an extinct group of animals related to arthropods. This exquisite example from the Chengjiang biota of China measures 8.2 cm (3¼ in) in length.

diagnostic for an extant phylum and fall outside the phylum as circumscribed by the living representatives of the phylum and their latest shared common ancestor.

Some Cambrian fossils are extremely peculiar, for instance *Opabinia*, a stem-group arthropod possessing five eyes and an elephant trunk-like proboscis emerging from the front end of the animal and terminating in a small claw. However, few can rival *Hallucigenia*, named for its bizarre and dream-like appearance. Fossils of *Hallucigenia*, which have been found at both Chengjiang and in the Burgess Shale, are up to 5.5 cm (2 1/6 in) long and comprise an elliptical head attached to a trunk-like body. One side of the body bears seven or eight pairs of appendages terminating in tiny claws, and on the opposite side are the same number of pointed spines. Walcott believed this fossil to be an annelid worm, but the palaeontologist and evolutionary biologist Simon Conway Morris had other ideas when, in 1977, he first coined the genus name *Hallucigenia* and reconstructed it as an immobile animal. He believed *Hallucigenia* to have lived supported above the muddy seafloor by its spines and used the clawed appendages to capture food and transport it to a mouth on the head. As no animals living today have such a mode of life, Conway Morris was perplexed by where *Hallucigenia* belonged on the tree of life. Further research showed *Hallucigenia* to have been reconstructed upside down – in fact, the appendages were locomotory legs and the spines were defensive structures on the top of the animal. As for its affinity, *Hallucigenia* belongs to an extinct group called the lobopodians, which are related to the living phyla Onychophora (velvet worms) and Tardigrada (water bears).

Radiodonta is an order of stem-group arthropods first recognized as late as 1996. Most radiodontids come from the Cambrian, although the order appears to have survived until the Early Devonian. The radiodontid *Anomalocaris*, like *Hallucigenia*, occurs in both the Burgess Shale and Chengjiang biotas. *Anomalocaris* was a large animal for its time, measuring almost 50 cm (19¾ in) in length. It was a free-swimming, apex predator and had a segmented body, fan-shaped tail, and a head with large, stalked eyes. Two long frontal appendages projected forwards from the head and were used for grasping prey. The mouth of *Anomalocaris*, as with other radiodontids, is surrounded by a disc-like structure formed of radial plates. Isolated examples of these 'oral cones' were originally misidentified as fossil jellyfish. Likewise, the first specimens to be described as *Anomalocaris* consisted of detached frontal appendages and were believed to be the abdomens of shrimp-like crustaceans. Subsequent findings of more complete specimens have elucidated the morphology of these impressive animals.

The fossil record shows that many different kinds of animals lived in the oceans of the Cambrian period, but does this point to a sudden explosion of life on Earth? Did multiple phyla appear in a short space of geological time during the Cambrian period, or are we looking at the first appearance in the fossil record of phyla with

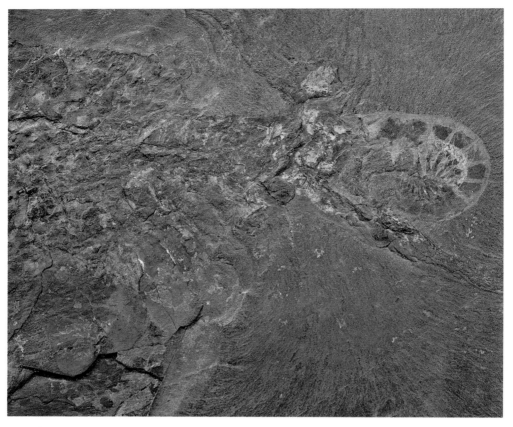

ABOVE The radiodontid *Anomalocaris* from the Burgess Shale of British Columbia, Canada was an apex predator in Cambrian seas. The two frontal appendages visible on the right of this image of a flattened fossil were used to grasp prey.

long and hidden prehistories? The latter could be explained by the evolution of biomineralized skeletons in phyla that had been soft-bodied during the Precambrian. However, fossils of soft-bodied organisms unequivocally belonging to extant phyla appear and proliferate in Cambrian Lagerstätten but are totally lacking in older rocks deposited in similar environments. In addition, trace fossils (burrows, trails etc., see Chapter 13) show a marked increase in abundance and variety in Cambrian rocks. Molecular clock studies have challenged a strict reading of the fossil record by estimating a strikingly earlier date for the origin of animals, approximately 800 million years ago. If correct, this would mean that animals evolved in pre-Cryogenian times and managed to survive the frigid 'Snowball Earth' environments of this tumultuous period.

'Molecular clocks' are routinely used nowadays to estimate the time of origin of biological groups. But what exactly are molecular clocks, and are they accurate timekeepers? When first mooted, the molecular clock hypothesis proposed that particular biomolecules – usually proteins, DNA and RNA – change at an approximately constant rate through time, i.e. the mutation rates are constant for the amino acids making up proteins and the nucleotide bases in DNA and RNA. Focusing on DNA, all living species have a 'genetic blueprint' defined by a unique sequence of nucleotide bases. The greater the difference between the sequences of two living species, the further back in time is the 'node', or most recent common ancestor, in a time-calibrated evolutionary tree. Molecular clocks require calibration. Although other methods are available, calibration is generally accomplished using the oldest occurrence of relevant fossils to estimate the dates of nodes on the tree. However, because the fossil record is incomplete, these dates will always be minimum ages, and various statistical models have been developed to estimate how much further back in time a group may have evolved. Once the molecular clock has been calibrated, it becomes possible to estimate the dates of all nodes on the evolutionary tree, and hence when each clade in the tree originated, including those that are unrepresented or poorly represented in the fossil record. Unfortunately, molecular clocks are fallible, and their veracity diminishes the further back in geological time they are applied. Variations in mutation rates (both through time and across the tree), uncertainties about molecular clock calibration points and other factors can lead to estimated dates of clade origin that are unrealistic when compared with data from groups that are common and abundant in the fossil record. Particularly troublesome is the large difference in the time of origin of animal phyla estimated using molecular clocks versus the fossil record – as remarked earlier, molecular clock estimates put this at over 200 million years earlier than the first appearance of these phyla as fossils at the time of the Cambrian Explosion.

3 Trilobites and other arthropods

There are more living species of arthropods than all other animal phyla put together. Most of this enormous diversity is accounted for by insects, especially beetles, but other types of arthropods which are extremely common today feature prominently in the fossil record, none more so than the extinct trilobites from the Palaeozoic. A defining feature of the phylum Arthropoda is the exoskeleton consisting of multiple elements, including articulated appendages. To grow, arthropods must periodically shed their exoskeletons in a moulting process called ecdysis, inflate their bodies, and manufacture a new exoskeleton. Aquatic arthropods respire using gills whereas terrestrial arthropods most often employ a system of branching tubes (tracheae) to carry oxygen to the tissues.

Arthropods were one of the first animal phyla to make an appearance in the fossil record, and studies of Cambrian Lagerstätten, such as the Burgess Shale of Canada and the Chengjiang deposits of China (see Chapter 2), have revealed a spectacular early diversification of arthropods in the sea at this time. Later in the Palaeozoic, various groups of arthropods colonized the land.

Radiodontids and a few arthropods have already been described or mentioned in Chapter 2. As for the rest, it would be impossible here to cover all groups of arthropods for which fossil examples are known. Instead, the focus is on a selection of fossil arthropods that are common, of palaeontological importance, or spectacular in appearance.

Trilobites

Trilobites are very much the movie-star fossils of Palaeozoic invertebrates. These endearing animals feature, for example, on the emblems of the Palaeontological Association and the Geologists' Association in the UK, as well as the Paleontological

OPPOSITE A fine example of the Silurian trilobite *Calymene blumenbachii*, the exoskeleton fully articulated. Measuring 4 cm (1½ in) in length, this fossil was collected from the famous Wenlock Limestone locality at Dudley in the West Midlands of England.

Society in the USA. Trilobites range through almost the entire Palaeozoic era and have been found on all continents of the globe. First appearing abruptly in rocks that are 521 million years old, in the early Cambrian period, they are last represented in the final stage of the Permian period some 252 million years ago. In total, therefore, trilobites as a group existed for about 277 million years of Earth history. Approximately 20,000 species of trilobites have been recognized, distributed between about 5,000 genera. Trilobites are a much-studied group and are used widely in stratigraphical correlation and for analyses of ancient environments and palaeogeography. In recent years, exquisitely preserved and rare trilobites have commanded high prices and have entered the investment portfolios of wealthy people. Fake trilobites (p. 14) have become commonplace, most are crudely executed, and some put together so imaginatively that they do not resemble any real species.

Trilobites owe their abundance as fossils to having an exoskeleton being partly reinforced with calcite, the calcium carbonate mineral that is highly stable through geological time. Only the upper, dorsal surface of the trilobite skeleton is calcified. The lower part, including limbs and antennae, consists only of organic chitin and is seldom preserved. As with other arthropods, the trilobite skeleton consists of numerous separate elements that were prone to fall apart after the death of the animal unless conserved by rapid burial. All trilobites had a head (or cephalon), thorax, and tail (or

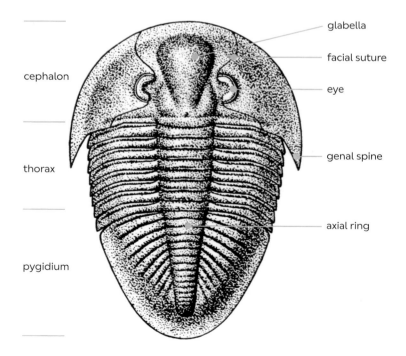

RIGHT *Basilicus*, a typical Ordovician trilobite with the main components of the skeleton labelled.

pygidium). These three components are often found separated, and the thorax can be disarticulated into its constituent segments. The threefold division of the skeleton into cephalon, thorax and pygidium is complemented by a threefold division across the width of the body from which the name trilobite originated. This comprises an axial lobe running along the midline of the animal, with lateral (or pleural) lobes on each side. Tubercles, spines, pits and ridges may adorn various parts of the exoskeleton. Each had a function, even though these are not always well understood.

Starting from the tail end, trilobite pygidia consist of a series of up to 30 fused segments. Often the pygidium is smaller than the cephalon but in some trilobite species it is large and shovel-like. The thorax consists of a variable number of segments, frequently a dozen or so, depending on the species; species with more segments tend to be geologically older. However, the number of thoracic segments is generally constant in adult trilobites belonging to a particular species. The cephalon is more complex. A lobe called the glabella, which might at first be mistaken for the brain of the animal, is actually a dome overlying the stomach. So-called facial sutures are present on either side of the glabella. These mark lines along which the skeleton split during moulting. The trilobite's pair of eyes are usually situated along the facial sutures. Backward-pointing genal spines often occur at the corners of the cephalon, and a shelf-like fringe may be developed around the outer border. Although less often observed, an important calcified part of the trilobite skeleton is the hypostome, a shield-shaped plate positioned beneath the glabella on the ventral side of the animal.

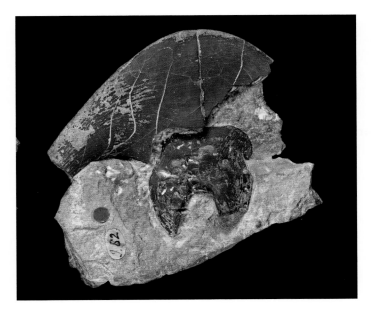

LEFT Underside of the head of the trilobite *Isotelus platycephalus*, showing the two-pronged hypostome which covered the mouth and had a role in feeding. Measuring 7 cm (2¾ in) across, this fossil was collected from the Ordovician Black River Group of St Joseph's Island, Ontario, Canada.

RIGHT Swedish specimen of the Cambrian Alum Shales Formation with a bedding plane completely covered by fossils of the tiny trilobite *Agnostus pisiformis*. The species name refers to the pea-like appearance of the fossil.

There are about 20 known species of trilobites with preservation of the limbs and antennae on the underside of the animals. These structures are not calcified, hence their rare preservation, which is often due to replacement by pyrite. Segments of the thorax and pygidium are each provided with a pair of limbs, and there are usually three pairs of limbs beneath the cephalon. Trilobite limbs are said to be biramous because they have two branches joined at the base. In at least some species the lower branch is a walking leg whereas the upper branch seems to form a gill, although in some species it could have functioned as a paddle for aeration or to facilitate swimming.

Most adult trilobites are 15 cm (6 in) or less in length, but depending on the species they range from under 1 cm to over 70 cm ($^2/_5$ to 27½ in). Some of the smallest species are found in the Cambrian–Ordovician order Agnostida. These trilobites have a short thorax with only two or three segments, and a cephalon and pygidium that are of similar size and shape. Mass accumulations comprising countless numbers of tiny Cambrian agnostids form limestones in the Alum Shale Formation of Scandinavia. At the other end of the size spectrum, the world's largest trilobite is *Isotelus rex* from the Ordovician of northern Canada. The biggest specimen of this 'king of the trilobites' measures 683 mm (27 in) and is incomplete, an estimated 40 mm (1½ in) of the pygidium having been lost to weathering before its discovery in 1998. To grow to this enormous size, *Isotelus rex* is estimated to have moulted 18 or 19 times during its lifetime.

Where and how did trilobites live? Until recently it has been assumed that trilobites inhabited only the salty waters of the sea, but new evidence suggests that a few species were able to migrate into the brackish waters of river estuaries. One thing is clear – trilobites never ventured onto the land surface, contrary to some science fiction movies. Detailed analysis of the preserved exoskeleton, coupled with knowledge of analogous living arthropods, has allowed a glimpse of the myriad of ecological niches occupied by trilobites in Palaeozoic oceans.

Trilobites typically had good eyesight. Much is known about their vision because the two eyes of trilobites have lenses made of crystalline calcite that has survived the hundreds of millions of years since the animals were alive. Like insects, trilobites have compound eyes. Three different kinds of eye have been distinguished: holochroal, schizochroal and abathochroal. By far the commonest are holochroal eyes – these are found in the earliest Cambrian trilobites as well as the last to become extinct at the end of the Permian. Holochroal eyes consist of hundreds of tightly packed hexagonal or circular lenses, each lens typically about one-tenth of a millimetre in diameter. The lenses are individual crystals of calcite with a precise crystallographic orientation, allowing a single beam of light to be focused on a photosensitive receptor, which is not fossilized. In the remarkable Ordovician species *Asaphus kowalewskii*, the eyes are at the ends of long stalks. Specialized schizochroal eyes are found in Ordovician to Devonian phacopine trilobites and have no known analogues among living animals. They have fewer, but larger, lenses of sophisticated design, which are up to three-quarters of a millimetre in diameter, convex and separated from one another. Light passing through the upper lens unit of calcite that sits in a bowl of different refractive index, which serves to improve the image received from the environment. The least common eye type in trilobites is the

BELOW Stalked eyes are a feature of the Ordovician trilobite *Asaphus kowalewskii* from the St Petersburg region of Russia. The eye stalks can be a couple of centimetres long in this species.

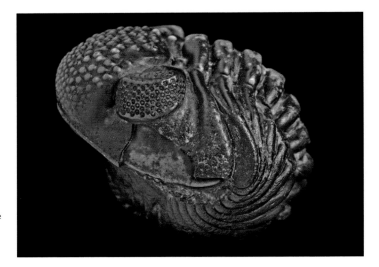

RIGHT Enrolled example of the Devonian trilobite *Eldredgeops rana* about 5 cm (2 in) in size. The schizochroal eyes of this phacopine trilobite from the Spanish Sahara contain a relatively small number of large lenses.

abathochroal eye, which is found only in Cambrian eodiscid trilobites. This resembles a holochroal eye but with lenses less regularly organized and spaced wider apart.

The majority of trilobites were inhabitants of the seabed, scuttling around along the surface of the sediment and sometimes leaving trails known by the trace fossil name *Cruziana* (p. 246). A few inhabited shallow burrows in the sediment. Three principal types of feeding have been suggested. Many trilobites were predators or scavengers with hypostomes buttressed or otherwise strengthened to cope with the need to handle bulky and tough items of food. Sediment-feeding trilobites were generally small and had free hypostomes that could possibly have been used as scoops. Finally, filter-feeding trilobites had long genal spines and a highly convex cephalon beneath which was situated a chamber where sediment was stirred up and food particles passed to the mouth. This feeding mode has been inferred for trinucleid trilobites characterized around the edge of the cephalon by numerous pits believed to represent exit routes for feeding currents. Among these filter feeders are a few lineages of trilobites that lost their eyes. Blindness was not quite the disadvantage it might at first seem for these trilobites as they were not visual hunters and many lived at depth in the ocean.

On at least half a dozen occasions, trilobites evolved to colonize the open oceans. These free-swimming trilobites, which are found in deep-water shales, had large bulbous eyes giving them a greater field of view than their relatives living on the seabed. A few of these pelagic trilobites could even see downwards, some probably inhabiting the zone of the oceans where light levels were very low, hence the need for large eyes to capture the small amount of light available.

LEFT Measuring 3 cm (1 in) in length, this specimen of the eyeless trilobite *Trinucleus* from the Welsh Ordovician shows the characteristic rows of pits around the edge of the head that may have been used in filter feeding.

BELOW A formidable array of defensive spines occurs in this 10 cm (4 in) long specimen of the trilobite *Comura* from the Devonian of Morocco.

As with all animals, trilobites fell victim to predators. Fossils are known of Cambrian arthropods with non-calcified skeletons containing fragments of trilobites in their guts. Numerous trilobites have been described as showing evidence of damage seemingly inflicted by biting predators, some surviving the attacks and making repairs to their exoskeletons. Trilobites show clear adaptations against predators. The most obvious are sharp, thorny spines seen, for example, in the Moroccan Devonian trilobite *Comura*. Like the modern woodlouse, many trilobites, particularly after the Cambrian, evolved the ability to roll themselves up into a tight ball, hiding their vulnerable underside. Enrolment could be very precise with the thorax flexed and the edges of the cephalon and pygidium fitting together to give a tight seal.

Two trilobites of cultural interest deserve to be mentioned. The first is *Calymene blumenbachii*, first described from the Silurian Wenlock Limestone Formation of Dudley in the English West Midlands. Fully articulated and enrolled examples are both common. First discovered in the mid-1700s, this trilobite is of medium size and has 13 thoracic segments. Specimens of *Calymene blumenbachii* found their way into the hands of early naturalists across Europe and became key players in debates that established the identity of trilobites as arthropods. A constant supply of specimens for study and ornamental purposes was ensured by active quarrying and mining of the Wenlock Limestone for use as a building stone, in lime manufacture and as a flux for the iron industry. Quarrymen referred to *Calymene blumenbachii* as the 'Dudley Locust', often supplementing their normal income by selling the specimens they found. Some versions of the Dudley Borough crest feature images of the Dudley

BELOW LEFT Perfect specimen of the Silurian trilobite *Calymene blumenbachii* set in the gold mount of a Victorian brooch.

BELOW RIGHT The small trilobite *Elrathia kingii* is extremely abundant in Cambrian rocks of the House Range in Utah, USA. This example is 2 cm (¾ in) long.

Locust, and in the collections of the Natural History Museum, London there is a specimen mounted in a gold brooch made in Victorian times.

A small trilobite called *Elrathia kingii* occurs in great profusion in the Cambrian Wheeler Formation of Utah – up to 500 individuals can be found on a square metre of bedding plane! This trilobite is thought to have inhabited parts of the seabed with low oxygen levels where it may have fed on sulphur bacteria, perhaps even hosting these microbes as symbionts in its own tissues. Extraction of disc-shaped specimens of *Elrathia kingii* resembling ancient coins is facilitated by the ubiquitous reinforcement of the thin exoskeletons by encrustations of inorganic calcite cement on their undersides. Dubbed America's most common trilobite, specimens of *Elrathia kingii* are collected for mounting in jewellery or as fridge magnets. In the past they were drilled and hung from necklaces made by the Pahvant Ute people who believed the trilobites could help to cure diphtheria and various other ailments.

Eurypterids

Eurypterids are uncommon fossils, but they more than make up for their rarity by being of spectacular size, none more so than *Jaekelopterus rhenaniae*. This gigantic Devonian eurypterid is known only from partial remains but is estimated to have reached 2.5 m (8 ft) in length, which would make it the largest known arthropod ever to have lived. Indeed, although adults of the smallest eurypterids measured just a few centimetres, almost one-third of genera included species likely to have grown to at least 80 cm (31½ in) in length.

The first eurypterids appeared in the Ordovician, were most numerous and varied in the Silurian, less so in the Devonian, and became extinct in the Permian. The common name for eurypterids is sea-scorpions, accurately reflecting their fearsome nature even though not all 250 described species lived in the sea, and their relationship with true scorpions remains a matter for debate. Eurypterids do, however, show various morphological similarities to scorpions. They have a shield-like head (prosoma) with two eyes, a tapering and segmented abdomen divided into pre-abdomen and post-abdomen, and a tail (telson) that can be pointed or flap-like. Six pairs of appendages are borne on the underside of the head, often including one pair at the tail end that are shaped like paddles for swimming. Most fossils consist of moults (discarded exoskeletons), with appendages often present. Unlike trilobites, eurypterids lacked a calcified skeleton – their protein-rich chitinous skeletons were thin but tough and are preserved in various stages of degradation to the element carbon.

Eurypterids have been found in rocks deposited in a range of environments, from deep to shallow marine, brackish and freshwater. Some eurypterids left trackways, made by

ABOVE The 'sea scorpion' *Eurypterus* has a pair of paddle-like appendages for swimming, seen clearly in this 12 cm (4¾ in) long fossil from the Silurian of New York State, USA.

the appendages, and drag marks, made by the tail, on sediment surfaces. It has long been suspected that a few eurypterids were amphibious, capable of emerging from the water onto the land. New research has shown the presence of organs resembling the 'book lungs' of spiders seemingly adapted to air breathing. One hypothesis has suggested that eurypterids behaved like modern horseshoe crabs, migrating towards the shoreline to moult and mate.

As for diet, at least three modes of feeding seem to have been adopted by eurypterids. Most had streamlined bodies and were fast-swimming predators able to pursue and capture their prey. This group declined from the Devonian onwards, possibly because of competition for food from fishes with jaws. A few may have been ambush predators, while some of the last eurypterids, such as *Hibbertopterus*, apparently used spiny appendages to rake the sediment, disturbing small animals that could then be eaten.

The Silurian Bertie Group of New York State and Ontario is famous for its eurypterids. Half a dozen eurypterid genera, including *Eurypterus* itself, occur in these sediments deposited close to the shoreline. Most of the eurypterid fossils represent moults and many became disarticulated before burial.

Crustaceans

Over 65,000 species of living crustaceans have been described. Some groups of these arthropods have skeletons reinforced by mineral layers and are better represented in the fossil record, including lobsters, crabs, barnacles, and the tiny ostracods which are dealt with among microfossils (see Chapter 12). The oldest fossil crustaceans date back to the Cambrian. Lacking mineralized skeletons, they are best known from Lagerstätten such as the Burgess Shale. Among the earliest crustaceans are the phyllocarids which have two valves enclosing most of the body.

Shrimps, lobsters and crabs are decapod crustaceans, a group defined by the possession of a thorax bearing five pairs of limbs. Decapods first appear as fossils in the Late Devonian but did not become particularly abundant as fossils until the Triassic. Unfortunately, their chances of being preserved as fossils are diminished by the fact that the mineralized layers of those decapods with calcified skeletons contain a large amount of organic material that is attractive to predators and decomposers. Nevertheless, claws and carapaces of lobsters and crabs are occasionally found in Mesozoic and Cenozoic limestones, some inhabiting reefs. However, the finest fossils occur within concretions in shales and clays. Phosphatic concretions in the Eocene London Clay Formation of southeast England have been a rich source of lobsters and crabs, many preserving the limbs along with the carapace. Bacterial decomposition of organic remains liberated phosphate from the buried remains, which was then precipitated around the decapods to form the hard concretions in which they are now encased. Parts of the decapods are visible on the outside of these concretions, but much careful preparation is required to reveal the fossils in their entirety. Elsewhere, burrowing decapods can be found fossilized in their burrows.

ABOVE *Nahecaris stuertzi* from the Hunsrück Slate of Germany is an Early Devonian phyllocarid crustacean. In this example the two shell-like valves enclosing the body have been flattened and splayed. The fossil is 7 cm (2¾ in) in width.

BELOW The two claws are very obvious in this lobster *Homarus morrisi* from the Early Eocene London Clay of Bognor Regis in West Sussex, England.

Lobsters appeared before crabs, which differ from lobsters in having a broad carapace beneath which is tucked the segmented abdomen. New evidence from gene sequencing has shown that crustaceans with crab-like features have evolved at least five times from lobsters, a process known as carcinization, beginning in the Early Jurassic.

Barnacles provoke painful memories for those who have had their bare feet lacerated when walking across rocks on the shoreline. The offending animals are acorn barnacles, or balanomorphs, with their sharp-edged, conical shells protecting the feathery appendages used to capture plankton when the tide comes in. Balanomorph shells consist of a basal plate cemented to the substrate, a ring of four to eight triangular rostral plates forming the sides of the cone, plus four opercular plates used to close the opening at the top when the tide is out. In contrast to the skeletons of other arthropods, barnacle plates are not moulted and typically contain tubular cavities. Opercular plates are often displaced or lost in fossil balanomorphs, and sometimes all that remains is the disc-like basal plate, still firmly cemented to the substrate.

Balanomorphs are unlike most other crustaceans because, when adult, they are permanently fixed to the substrate. Their fossil record commences in the Late Cretaceous but they did not become common until the Miocene, heralding what Charles Darwin, an aficionado of these crustaceans, referred to as the 'Age of Barnacles'

BELOW Acorn barnacle and a mass of dispersed barnacle plates photographed in the field at Castlepoint in the Wairarapa region of New Zealand, where the Early Pleistocene Castlepoint Formation outcrops in coastal cliffs.

with more than 2,100 living species. Clusters of fossil balanomorph barnacles can be found covering mollusc shells, and dense communities of these animals furnished prodigious quantities of plates, which accumulated on the surrounding seafloor, on occasions forming barnacle limestones, as in the Pliocene and Pleistocene of New Zealand.

Balanomorphs are by no means the most ancient barnacles. The oldest known fossil barnacles from the Carboniferous pre-date them by some 250 million years and had stalks like the gooseneck barnacles that can be seen today attached to driftwood and which are occasionally found as fully articulated fossils. These and other early species are notable in having plates composed of calcium phosphate rather than the calcite plates seen in all modern barnacles, the switchover in mineralogy first occurring in the Middle Jurassic.

ABOVE Gooseneck barnacle *Stramentum* from the Late Cretaceous Chalk of Kent, England. A couple of centimetres in length, it is uncommon to find these barnacles so completely articulated.

Insects and spiders

Whereas the arthropods described above are overwhelmingly inhabitants of the sea, insects and spiders are terrestrial animals but with origins that can be traced back to aquatic ancestors. In an evolutionary sense, insects are nested within a larger group known as Pancrustacea – they are essentially crustaceans that have evolved to live on the land. Astonishingly, almost a million living species of insects have been described and a further 4 million are predicted to await discovery. It would not be unreasonable to speculate that more than a hundred million species of insects have populated our planet since their first appearance in the Devonian. The distribution of insect fossils is patchy, although they are by no means uncommon and include representatives of all 30 or so orders of insects that are alive today. Although nearly 30,000 fossil insect species have been recorded, this must be just a tiny fraction of the insects that have existed, the lack of biomineralized hard parts making their fossilization unlikely. Insect cuticles, which are complexes of chitin and proteins, can decay rapidly. Even cuticles that become buried to depths where biological activity is minimal may undergo destructive chemical changes through geological time and finish up as solid organic matter known as kerogen. In general, insects with thicker cuticles, such as cockroaches and beetles, are more often fossilized than the delicate ones, such as butterflies and moths.

BELOW Wing of a giant Late Carboniferous insect from Derbyshire, England. Despite being fractured into several pieces, this type specimen of the 'Bolsover Dragonfly' *Tupus diluculum* is nevertheless impressive. The rock measures 22 cm (8 in) across.

Insects, along with three lesser-known groups (coneheads, springtails and two-pronged bristletails) form a group of arthropods called the Hexapoda, named in reference to their six legs. They have bodies comprising head, thorax and abdomen, compound eyes, and a pair of antennae. Most insects have two pairs of wings; in beetles, the front pair are transformed into hard cases (elytra) that cover the rear pair. However, the most primitive insects, first recorded in the Devonian Rhynie Chert of Aberdeenshire, Scotland, were wingless, and the evolution of wings in the Carboniferous triggered a boom in insect evolution. Their diversity has increased steadily since then, declining with the end-Permian mass extinction but relatively unperturbed by the end-Cretaceous mass extinction. Particularly notable among Palaeozoic insects is the Permian dragonfly-like genus *Meganeuropsis*, represented by a 33-cm-long (13-in) wing found in Kansas, USA, which makes *Meganeuropsis* the largest known insect ever to have lived. Similar gigantic insects have also been found in Late Carboniferous rocks, and it has been suggested that the dense, oxygen-rich atmosphere of the time may have allowed insects larger than any recent species to evolve and flourish.

Fine-grained sediments deposited in tranquil lakes or shallow seas close to the coastline and undisturbed by burrowing animals are most likely to preserve fossil insects. For example, the famous Bavarian Solnhofen Limestone, host to the earliest known bird *Archaeopteryx* (see Chapter 9), is a laminated Jurassic sediment formed in lagoons with limited connections to the sea. During periodic drying, insects, including spectacular large dragonflies, became caught in the sticky mud. Elsewhere, the Eocene Bembridge Marls on the Isle of Wight, UK contain a celebrated fauna of insects numbering some 200 species. Deposition occurred in brackish water adjacent to a subtropical forest inhabited by insects of various kinds but dominated in numbers

LEFT An exquisite specimen of the dragonfly *Turanophlebia* from the Late Jurassic Solnhofen Limestone of southern Germany, which is famed for its fossil fauna of animals that came to grief in a salty lake. The specimen is about 10 cm (4 in) across.

BELOW The butterfly *Lithopsyche antiqua* is known only from this one broken specimen from the Late Eocene Insect Limestone of the Isle of Wight, England. The wing on the left shows bands of preserved colour. The green spot on the left is about 3 mm in diameter.

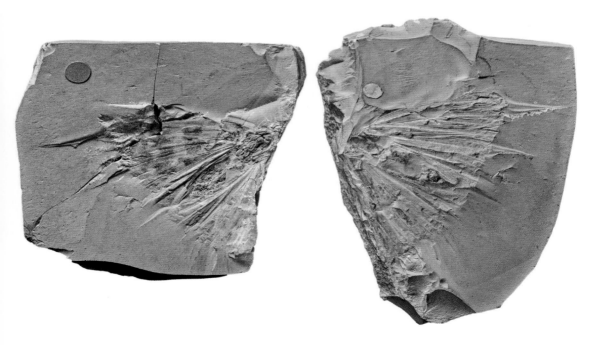

of individuals by flies, hymenopterans (wasps and their relatives) and beetles. As is so often the case, most of the fossils are represented by wings. The patterning of veins on the wings is sufficiently distinctive to be used by palaeoentomologists for identification purposes.

The semiprecious stone amber is famous for containing fossil insects that had the misfortune of being trapped in resin produced by ancient trees to fill lesions in the bark. Over time, the resin hardened by losing its volatile oils and polymerizing into copal, then darkening in colour through further polymerization and becoming amber. Hundreds of amber deposits are known worldwide. The oldest amber dates from the Carboniferous period, but amber containing fossil insects is rare before the Early Cretaceous. Baltic amber of Eocene age has encased huge numbers of insects, as well as spiders and occasional snails and small lizards. In total, more than 3,000 species have been described from Baltic amber, of which flies are the commonest

ABOVE Long-legged fly of the family Dolichopodidae preserved in Baltic amber of Late Eocene age. The fly is 2.3 mm long.

RIGHT Pendant made from Baltic amber containing two spiders, a harvestman (*Phalaphium*) and a lynx spider (*Oxyopes*). The pendant is 5 cm (2 in) long excluding the ring.

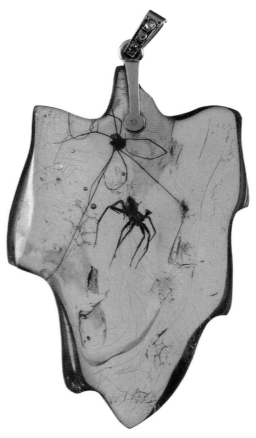

ABOVE Part and counterpart of the Late Carboniferous whip spider *Graeophonus anglicus* preserved in a small concretion from the Coal Measures of Coseley in Staffordshire, England.

insects, followed by hymenopterans. Most insects encapsulated in Baltic and other ambers are preserved as hollow moulds: little or nothing remains of the internal tissues. Although recent research has reported mineralized trachea and other soft tissues in Cretaceous Burmese amber, recovering dinosaurian DNA from the blood contained in the stomachs of mosquitoes in amber – the starting point for the re-creation of living dinosaurs in Michael Crichton's novel *Jurassic Park* – is not yet on the horizon.

Spiders are another diverse group of terrestrial arthropods. All are predators. They have eight legs, venom-injecting fangs and spinnerets from which silk is exuded for purposes that include web building. The head and thorax are fused to form a prosoma as in other chelicerate arthropods, a group to which scorpions, horseshoe crabs and eurypterids also belong. Fossils of spiders have been found back to the Late Carboniferous. However, although more than 1,100 species have been described, some preserved in exquisite detail, spider fossils are rather rare, especially if we discount those found in amber. This is not surprising in view of their delicate bodies and lack of mineralized hard parts. Notable are spectacular examples of trigonotarbids and whip spiders found in concretions in the Carboniferous Coal Measures. The former are extinct relatives of true spiders but lack spinnerets and so were unable to produce silk.

4 Ammonites and their relatives

Ammonites

Ammonites are the most iconic of all fossils. They have a universal appeal. For geologists, ammonites are extremely useful in correlating rocks of equivalent age. But this is not the reason that ammonites are so well known – because even the youngest children can instantly spot an ammonite, they are often the first fossils they collect. Older fossil enthusiasts love them for their handsome spiral shells, while people with little interest in fossils like wearing them in jewellery or displaying them as ornaments. The logarithmic spiral shells of ammonites have been an inspiration to artists and designers alike. Look around and you will see ammonite-based designs in drawings, ceramics and jewellery. They have even influenced architecture – some neoclassical buildings in southern England constructed in the early nineteenth century have capitals modelled on ammonites, including the house in Lewes that was the home of the early dinosaur expert Gideon Mantell (1790–1852). Ammonites also feature prominently in fossil folklore and were once believed to be coiled snakes miraculously turned to stone. St Hilda was supposed to have performed this feat, which allegedly explained the presence of ammonites on the seashore at Whitby where she was the abbess. Images of the saint are often accompanied by ammonites, as in the arms of St Hilda's College, Oxford. The idea that ammonites were petrified snakes was given impetus by unscrupulous dealers who sometimes carved snake heads onto the ends of the coiled shells, thereby increasing the monetary value of these 'snakestones'.

Ammonites are a long-extinct group of animals. To understand how they lived, we need to examine their fossilized shells, along with the rarely preserved soft parts, and draw analogies with the living animals to which they are closely related. Ammonites belong to a class of molluscs called cephalopods. Modern cephalopods include squid, cuttlefish and octopus. With a few notable exceptions, these extant

OPPOSITE 'Snakestone' made by carving a snake's head complete with eyes onto a fossil of the Early Jurassic ammonite *Dactylioceras commune* from North Yorkshire, England. 6 cm (2¼ in) across.

animals are poorly represented in the fossil record on account of their feeble or non-existent shells. On the other hand, many older groups of cephalopods had robust shells of calcium carbonate that fossilized well. The single survivor of these more ancient, shell-bearing cephalopods is the pearly nautilus, belonging to the subclass Nautiloidea (informally 'nautiloids'). Ammonites (order Ammonitida) themselves are placed in another superorder – Ammonoida ('ammonoids') –with several older orders of which the Goniatitida ('goniatites') and Ceratitida ('ceratites') are the most notable. The bullet-shaped internal skeletons of belemnites (order Belemnitida) are another kind of extinct cephalopod that are abundant as fossils.

Certain features unite these diverse types of cephalopods, living and fossil. All are active and intelligent animals with well-developed nervous systems. Octopuses are capable of problem-solving and are justifiably regarded as the 'brain-boxes' of the invertebrate world. They have advanced eyes resembling those of vertebrates, although independently evolved. Many cephalopods are capable of rapid swimming using a form of jet propulsion – water is sucked into the body and expelled rapidly

BELOW LEFT Unembellished specimen of the Early Jurassic ammonite *Dactylioceras commune*, measuring 9.5 cm (3¾ in) in width, from North Yorkshire, England. Covered by strong ribs, the shell of this species is loosely coiled or evolute.

BELOW RIGHT In the Early Cretaceous heteromorph ammonite *Aegocrioceras quadratus* the shell is partly uncoiled with successive whorls not in contact. This fossil is 12 cm (4¾ in) across.

through a funnel by the contraction of muscles, propelling the animal in the opposite direction through the water. Ink sacs present in some cephalopods are used to create a cloud of black melanin particles, producing a sort of smoke screen and confusing would-be predators. There are two kinds of appendages in cephalopods – arms and tentacles – the latter generally longer and capable of extension. The number of arms and tentacles varies greatly among different groups of cephalopods. Both arms and tentacles can be equipped with suckers, and the arms of some cephalopods also have hooks. Primitively, cephalopods all possessed multichambered shells, which are particularly clear to see in sections of ammonites. By adjusting the levels of liquid and gas in these chambers, the animals could control their buoyancy and hence the level at which they floated in the water. In cephalopods lacking shells, buoyancy is controlled chemically by ammonia or oils in the tissues. All cephalopods inhabit the sea; they are not found in rivers or freshwater lakes. If you find a fossil cephalopod

ABOVE Resembling a gastropod shell in its helicospiral coiling, this 6 cm (2¼ in) long fragment of the Late Cretaceous ammonite *Turrilites costatus* came from Rouen in northern France.

BELOW The shell of the Early Cretaceous heteromorph ammonite *Hamites* is initially coiled before becoming straight during later growth. About 15 cm (6 in) in length.

of any sort you can be sure that the rocks from which it came were deposited in a marine environment.

Ammonites usually have shells that are coiled spirally in a flat plane with an aperture at the end through which the 10 or so appendages and other parts of the body were extended. The successive whorls of the spiral overlap to varying degrees: in some species the overlap is slight, and the ammonite is said to be 'evolute'; in other species, later whorls almost entirely cover earlier ones, a condition known as 'involute'. There are, however, ammonites in which the spiral whorls do not touch at all, others with almost straight shells, and a few with corkscrew-shaped shells resembling snails.

Complete ammonites are typically between 5 and 20 cm (2–8 in) in diameter but there are a few much larger species. The largest ammonite ever found measured a whopping 1.74 m (5½ ft) across – bigger than the average height of a woman in the UK in 2023, which the Office for National Statistics puts at 1.64 m (5¼ ft)! But as the shell of this specimen of *Parapuzosia seppenradensis* from the Late Cretaceous of Westphalia, Germany was broken, it may have been as much as 2.5 m (8 ft) in

BELOW LEFT Sectioned specimen of the Middle Jurassic ammonite *Brasilia bradfordensis* from southern England. Septa subdivide the shell into chambers some of which are infilled with hardened sediment while others are lined by calcite crystals. The fossil is 16 cm (6¼ in) across.

BELOW RIGHT Polishing of this specimen of the large Middle Jurassic ammonite *Parkinsonia dorsetensis* has exposed the intricate suture lines where the septa meet the inner surface of the shell. 26 cm (10¼ in) across.

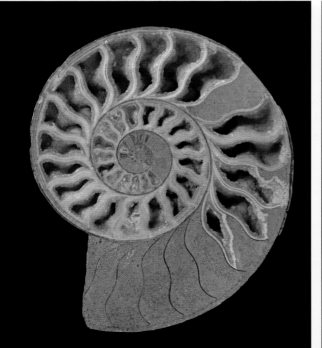

diameter when complete. Overall shell shape varies enormously between the 10,000 or more different species of ammonites. Some have plump, almost spherical shells, whereas the shells of others are flat and disc-shaped. In a few ammonites, the shell has a smooth surface but in most there are fine growth lines or regularly spaced ribs. The ribs are sometimes associated with nodes or occasionally spines, especially close to the outside of the whorls where some species also develop a ridge or keel running along the outer margin of the shell.

All ammonites have chambered shells. The chambers, which are separated from one another by partitions called septa, are best seen in sectioned specimens. The final chamber, closest to the aperture, is known as the body chamber as this is where the main part of the body of the animal was located during life. Older chambers, which constitute the phragmocone, would during life have contained the fluids used to regulate buoyancy. During fossilization, the chambers typically became fully or partly infilled with minerals crystallized from percolating solutions, hardened sediment, or a mixture of minerals and sediment. Passing through a hole in each of the septa is a tube known as the siphuncle. This is positioned near the outer sides of the whorls in ammonites and can be visible in sectioned specimens that have been cut through the exact mid-plane of the shell.

A defining characteristic of ammonites is the presence of the remarkably intricate traces formed where the sutures meet the whorls of the shell. These 'suture lines' are most evident in specimens in which the shell has been partly broken away or is lacking altogether, for example in ammonites preserved as internal moulds. The suture lines of different species are as unique to these species as human fingerprints and can be very useful in identification. Palaeontologists have long tried to explain why ammonites have such complex sutures. A popular theory has been that the shell was strengthened by its lengthy contact with the suture acting like a buttress. This would lessen the likelihood of the chambers imploding as the animal ventured into deeper waters where the pressure acting on the shell would have been high. However, new research has cast doubt on this idea – mathematical modelling of shell strength has failed to find any significant effect of sutural complexity. Furthermore, most ammonites seem to have inhabited water depths shallower than 250 m (820 ft), whereas the modern nautilus which has simple sutures has been observed swimming down to a depth of 700 m (2,300 ft). An alternative suggestion is that the large surface area of the complex sutures of ammonites had a role in improving buoyancy regulation. New research has shown that complex sutures may have made it harder for predators to puncture the shell. The jury is out.

Variations in all these features of the shell undoubtedly reflect differences in the behaviours and ecologies of ammonite species. Some ammonites probably escaped from predators by swimming quickly, employing the jet propulsion method common to cephalopods, whereas others relied on their armoured shells for passive defence. It is likely that the fastest-swimming ammonites had streamlined shells that were narrow and smooth, whereas those with rotund shells ornamented with coarse ribs or nodes were more sluggish. Ammonites with irregularly coiled shells, known as heteromorphs, may also have been slow-moving, perhaps feeding on plankton rather than larger more active prey. Fossil assemblages quite commonly consist of an assortment of different ammonite species, which may have swum at different depths, some feeding near the seafloor, others catching prey at higher levels in the water column.

A few ammonite species show conspicuous differences between males and females. Shells belonging to males are smaller and known as microconchs, those of the larger females are called macroconchs. Early growth stages of microconchs and macroconchs are identical, but as sexual maturity was reached differences began to develop, not just in shell size but also in shape, with the males of some species developing curved extensions called lappets at the aperture.

Ammonites built their shells entirely from aragonite, a calcium carbonate mineral that is prone to being dissolved through geological time. Consequently, a lot of

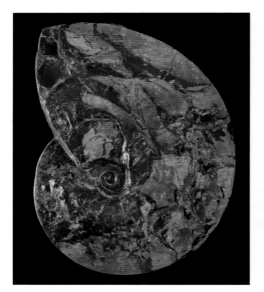 

ABOVE Spectacular iridescent shell of the Cretaceous ammonite *Placenticeras* from Alberta, Canada. The coiling of the 20 cm (7¾ in) wide shell is involute, with successive whorls strongly overlapping.

ABOVE RIGHT Aptychus consisting of a pair of calcite plates located near the aperture of a poorly defined shell of the ammonite *Oppelia* from the Late Jurassic Solnhofen Limestone of Germany. The shell measures 5.4 cm (2¼ in) in diameter.

LEFT Marston Marble is an Early Jurassic limestone from Somerset in England packed with ammonites. Each shell in this view is just a few centimetres in diameter.

ammonites are preserved as internal moulds (steinkerns) consisting of hardened sediment, the shell itself having been lost. In others, the shell is replaced by other minerals such as calcite or gold-coloured pyrite. Ammonites that do have the original aragonite shell intact can be exquisite in appearance. These shells preserve the nacre, otherwise known as mother-of-pearl, which is sometimes lustrous and white resembling porcelain, but more often it consists of iridescent bands spanning the colour spectrum from yellow, through brown and green, to blue and purple. Fine examples of nacre can be seen in thick-shelled specimens of the Cretaceous ammonite *Placenticeras*. (Commercially mined in Alberta, Canada, pieces of these shells are cut and polished to be used as a gemstone in jewellery under the name ammolite). At the other end of the spectrum of preservation are indistinct outlines of ammonite moulds visible in limestones where all aragonite shells have been dissolved away. Such is the case in the famous German deposit called the Solnhofen Limestone, host to such palaeontological wonders as the earliest known bird *Archaeopteryx*. Here, however, pristine examples can be found of another part of the ammonite skeleton, the aptychus. Unlike the coiled shells, aptychi are made of calcite, which is more resistant to dissolution. The ammonite aptychus, comprising a pair of plates abutting along a straight edge, probably formed part of the jaw apparatus of the living animal.

Some of the finest ammonites are found inside hard concretions present in shales and mudstones. Although ammonites in the surrounding sediment may be crushed, the growth of the cement forming the concretion protected the ammonite from compaction. Small concretions may encase single ammonites, whereas larger concretions can host several ammonites. Dense accumulations of ammonite shells are occasionally found. These may be due to slow rates of sediment deposition, with ammonite shells raining down onto the seafloor over a long period of time. In this case, the shells are commonly eroded or corroded and encrusted by other marine organisms. In contrast, some accumulations of pristine ammonites possibly represent mass mortality events, perhaps following spawning, as happens among some cephalopods living today. A spectacular example is the ammonite 'graveyard' of the Early Jurassic deposit from Somerset known as the Marston Marble.

True ammonites are found in Jurassic and Cretaceous rocks worldwide. They evolved rapidly, each species having a relatively short duration, which makes them extremely good fossils for determining the relative ages of rocks. Around 70 ammonite time-zones have been distinguished in the Jurassic of Europe. Given that the Jurassic period lasted for 56 million years, each time-zone had an average duration of about 800,000 years. To give an example, the Bifrons Zone is defined by the occurrence of an ammonite species called *Hildoceras bifrons*, the genus named for the aforementioned seventh century St Hilda, abbess of the monastery at Whitby in North Yorkshire where this ammonite is especially common. The same species has also been found in North Africa and elsewhere in Europe in rocks that can be reasoned to have been deposited at the same time as those in Whitby.

After flourishing in the world's oceans for 135 million years, the last ammonites disappeared 66 million years ago at the same time as dinosaurs and many other animals and plants. This end-Cretaceous mass extinction event is generally attributed to the catastrophic effects of an asteroid impacting the Earth (see p. 26). Collapse of the plankton that were a direct or indirect source of food for ammonites could have brought about the demise of these iconic animals.

Ceratites, goniatites and nautiloids

Chambered shells coiled into spirals characterize some other groups of cephalopods. These include predecessors of the true ammonites, such as the ceratites and the goniatites, both of which are ammonoids. Superficially, these two ammonoid groups can look almost identical to ammonites, but neither group has the complex suture lines seen in the ammonites of the Jurassic and Cretaceous. The long-ranging, extant nautiloids differ from ammonoids in several ways, most particularly in having siphuncles passing through the centre of the chambers rather than being located at the outer edge (see front cover).

ABOVE Ceratite ammonoid, *Ceratites nodosus,* from the Triassic of Germany, measuring 6 cm (2¼ in) across. The distinctive shapes of the suture lines have been accentuated in this specimen by painting successive chambers black and white.

ABOVE RIGHT Early Carboniferous goniatite ammonoid with an involute shell and simple, zig-zag suture lines. The shell is 4.5 cm (1¾ in) across.

In ceratites, the sutures consist of a series of complex, frill-bearing lobes facing the aperture, with simple saddles in between. Ceratites are typical of the Triassic, although some species are known from the Permian. The most intensively studied examples come from the Muschelkalk (mussel limestone) quarried as building stones and for cement manufacture in central and western Europe, especially Germany. Muschelkalk ceratites are generally preserved as internal moulds left after the chambers of the shells had become filled by lime mud and the shell lost through dissolution. Commonly, not all the chambers of the phragmocone were completely infilled by sediment, which had to enter from the body chamber via the narrow siphuncle linking the older chambers forming the phragmocone, or by breakage of the shell to expose the chambers. Despite the imperfect preservation of many Muschelkalk ceratites, palaeontologists have been able to discern a detailed evolutionary sequence of species in the Upper Muschelkalk of the German Basin. This has allowed the recognition of 14 zones averaging 250,000 years in duration. Triassic ceratite ammonoids must have evolved at least as rapidly as the ammonites that succeeded them in the Jurassic.

Goniatite sutures are even more simple than those of ceratites, lacking frills and generally forming zig-zag patterns. This group of ammonoids can be found in Devonian, Carboniferous and Permian rocks. The shells of goniatites tend to be smooth or adorned with low ribs or growth lines. Although some Late Devonian species reached 80–100 cm (31½–39½ in) in diameter, a good many are pea- or marble-sized. Polished Moroccan goniatites in dark-coloured limestone are a frequent sight in shops selling fossils or decorative items. As with ammonites and ceratites, goniatites have proved to be useful in stratigraphy. This includes goniatites found in bands of marine limestone or mudstone intercalated with non-marine sandstones and shales in the mid-Carboniferous rocks of the British Isles, for example at Slieve Anierin in Eire. The non-marine sediments are difficult to date but the goniatites in the marine bands allow crucial correlations to be made with rocks as far distant as North America.

Whereas ammonoids have been extinct for some 66 million years, nautiloids are represented in modern seas by nine species of *Nautilus* and two of *Allonautilus*. Not only have they out-lasted ammonoids, but nautiloids were the first cephalopod molluscs to evolve, appearing in the Cambrian. Nautiloids of Mesozoic and Cenozoic age tend to resemble living species and have tightly coiled broad shells. However, Palaeozoic nautiloids can be very different, many having straight shells. Spectacular examples of these so-called 'orthocones' can be seen in paving stones, for example at Hampton Court Palace,

ABOVE Sections of straight, orthocone nautiloids are often to be observed in paving stones used for the floors of historical buildings in Britain. These usually come from Ordovician rocks quarried on the Swedish island of Öland.

RIGHT Sectioned shell of the nautiloid *Cenoceras astacoides*. The siphuncle can be seen passing through the centres of the chambers in this 13 cm (5 in) wide fossil from the Early Jurassic of Whitby, England.

FAR RIGHT Egg-shaped shell of *Gomphoceras pyriforme*, a peculiar Silurian nautiloid that can grow up to 11 cm (4¼ in) in length.

west of London, UK. Quarried in Sweden or Russia, these slabs of red Orthoceras Limestone reveal sectioned shells typically 3 or 4 cm (1–1½ in) wide and a few tens of centimetres long, expanding gradually towards the aperture. Gently curved septa, convex in the direction of the narrower end, subdivide these Ordovician fossils into the multiple chambers of the phragmocone, with the longer body chamber sometimes preserved at the broader end. The largest orthocone shell is an Ordovician species from New York State appropriately called *Endoceras giganteum*, possibly measuring as much as 10 m (33 ft) in length. These enormous animals must have been formidable predators in Ordovician oceans before jawed fish had evolved.

Other Palaeozoic nautiloids have shells that are at first spirally coiled before straightening out, a form that was to reappear later in heteromorph ammonites. Among the most peculiar nautiloids are those with egg-shaped shells, such as the Silurian genus *Gomphoceras*. About half to two-thirds of the shell comprises the body chamber, there are 10 or fewer chambers in the phragmocone, and the aperture is curiously constricted. Unlike the large orthocones, many of which would have been fast swimming predators holding their shells more-or-less horizontally, *Gomphoceras* is likely to have been a more sluggish animal oriented vertically in the water with the aperture facing downwards.

Belemnites

In contrast to ammonoids and nautiloids with their external shells, belemnites had internal shells totally enclosed by the soft body when alive, and superficially resembled modern squid and cuttlefish. First appearing in the Late Triassic, belemnites became extremely common in the Jurassic and Cretaceous, particularly in colder water regions of the Mesozoic world, inhabiting shallow seas probably no deeper than 200 m (656 ft). By far the commonest parts of belemnites to be fossilized are the so-called guards, otherwise known as rostra. These bullet-shaped objects range in length from a few centimetres in the smallest species to almost 80 cm (31 in) in the largest. Some of the longest are found on the coast of the Isle of Skye. Annoyingly for the fossil collector, belemnite guards are apt to break into pieces. This does, however, reveal the internal structure of the guard, consisting of coarse calcite crystals radiating from the centre. Like the rings of a tree, growth lines visible in broken or sectioned specimens demonstrate how the guards grew.

Belemnite guards are so dissimilar to the skeletons of any animals living today that early naturalists were puzzled about their origin. Some believed them to be inorganic structures. There was even a widespread belief that belemnites formed during thunderstorms, being cast down to the ground like missiles. This led to belemnite guards being called 'thunderbolts' in folklore. Only with the discovery of the less commonly

ABOVE This guard of the Early Jurassic belemnite *Parapassaloteuthis robusta* from Whitby, England has been sectioned to reveal the growth lines as well as the V-shaped cavity (alveolus) into which the chambered shell was slotted.

preserved parts of the belemnite skeleton did their identity as cephalopod molluscs become clear. A conical cavity is present in the broad end of the belemnite guard. In well-preserved specimens, a chambered shell is slotted into this cavity and extends someway beyond it. This structure is the equivalent of the phragmocone of ammonoids and nautiloids. It would have functioned in controlling the buoyancy of the living animal through adjustments in the fluid levels in the chambers. The reason that belemnite phragmocones are preserved far less often than guards, aside from their greater fragility, is that they are composed of the more soluble mineral aragonite, in contrast to the geologically stable calcite of the guard. Thus, like ammonites with their aragonite shells and calcite aptychi, belemnites used two different minerals to construct different parts of their skeletons.

Some belemnites have been discovered with preserved soft parts. One example of *Hibolithes* from the Late Jurassic of Germany had 10 arms, each equipped with 40 small hooks (a careful search through rocks containing belemnites can reveal scattered dark-coloured hooks generally less than 5 mm long). The German specimen of *Hibolithes* is also preserved an ink sac, confirming that belemnites were capable of releasing ink like many recent cephalopods to escape from predators. Indeed, belemnites have been reconstructed to resemble a modern squid in external appearance. Internally, however, they were very different – no living squids have a guard, and only the peculiar extant genus *Spirula* possesses a chambered shell of calcium carbonate. The belemnite guard can be interpreted as a counterbalance to the weight of the main body of the animal, allowing a horizontal orientation to be maintained in the water. Taking into account the phragmocone and soft body including tentacles, the largest belemnites with guards 80 cm (31 in) long would have measured close to 3 m (almost 10 ft) in length.

Dense aggregations of belemnite guards are sometimes found strewn over bedding planes. Referred to as 'belemnite battlefields', the origin of these concentrations is uncertain, although they were almost certainly not masses of warring belemnites as this name implies. Some belemnite battlefields may have been the result of regurgitation by large predators of these indigestible, bullet-shaped chunks of calcite. Prominent among the predators of belemnites were ichthyosaurs, the large dolphin-shaped swimming reptiles that populated Mesozoic seas (p. 170). Although the guards of the

FAR LEFT This 22.5 cm (8¾ in) long specimen of the Jurassic belemnite *Cylindroteuthis* preserves a crushed phragmocone attached to the guard.

LEFT Five fossils of the belemnite *Belemnitella minor* from the Late Cretaceous Chalk of Norfolk, England. The guards are oriented as though they were projectiles flung down from the sky, corresponding to the folklore that belemnites were thunderbolts.

belemnites may have been regurgitated by their ichthyosaur predators, other parts of the prey animals were not, particularly the small arm hooks. These hooks often feature among the remains of the last meals eaten by ichthyosaurs, preserved as a mass of material where the stomach was located.

Along with ammonites, belemnites perished during the catastrophic events at the end of the Cretaceous period. The great majority of cephalopods living today, including most squid and octopus, lack chambered shells of calcium carbonate. Ancestors of modern cephalopods possessed shells but these were lost or strongly reduced during evolution. The buoyancy function of the chambered shell is replaced in squid by an organ comparable to the swim bladder of fishes. By adjusting the levels of ammonium and oils in this organ, a squid is able to ascend or descend in the water. With the loss of their calcified shells, cephalopods became a minor component of the fossil record – fossil cephalopods are far less common in the Cenozoic than they are in the Palaeozoic or Mesozoic. There are more than 300 species of squids living today and a roughly equal number of octopus, not to mention species belonging to two other groups of cephalopods, cuttlefish and the vampire squids. Yet these diverse animals have not left much trace in the fossil record for the last 66 million years of geological time. It is one thing to exist, quite another to be fossilized.

5 Superabundant shells

Shells are the bread-and-butter of the fossil record and are likely to form the basis of most amateur fossil collections. They are easy to find, as they occur in almost all macrofossil-bearing Phanerozoic sedimentary rocks deposited in the sea or in lakes during the last 539 million years, and they come in many shapes and sizes. The great thing for fossil enthusiasts is that shells are often present in prodigious numbers. Sometimes they are scattered through the sediment but in other instances they are concentrated in shell beds, distinct layers packed with shells. Shell beds may contain shells of a single or numerous species and originate in different ways; for example, as current-swept accumulations, through winnowing of the surrounding sediment to leave only the shells, or simply because the shell-bearing animals lived in dense populations. Many fossil shells shine and would not look out of place among contemporary shells found on a modern beach, while others are more obviously fossils and are quite unlike any living species. Whole and broken fossil shells can be observed in urban settings all around the world as they are frequently the main constituents of the limestones used as building stones.

This chapter describes the shelled fossils belonging to two invertebrate phyla: Brachiopoda and Mollusca (apart from the Class Cephalopoda, which were covered in the previous chapter). Both are commonly found in rocks formed over the past 500 million years. The lesser-known brachiopods are particularly abundant in Palaeozoic deposits, while molluscs tend to be more numerous and diverse in Mesozoic and Cenozoic deposits.

Bivalve molluscs

The Bivalvia is a diverse class of molluscs comprising over 9,000 extant species. The great majority inhabit the sea, but some populate freshwater lakes or rivers. Most of

OPPOSITE Shell of the left-handed, or sinistral, gastropod *Neptunea angulata*. The aperture is on the left in contrast to the great majority of gastropods which are dextral and have the aperture on the right. This specimen is from the Pleistocene Red Crag of Suffolk, England and measures 15 cm (6 in) in height.

us can easily recognize at least some types of bivalves, which routinely turn up in shell collections as well as on dinner plates, including oysters, mussels, scallops, cockles and clams. The edible soft parts are enclosed by two articulated valves that are made of calcium carbonate, explaining the great abundance of fossil bivalves. Much can be deduced about the lifestyles of extinct bivalves from the shapes and structures of these shells, shedding light on the ancient environments they inhabited.

Sometimes referred to as 'pelecypods' or 'lamellibranchs', bivalves range in size from less than a millimetre to more than a metre in the case of the living clam *Tridacna* and even larger in some Cretaceous inoceramid bivalves (p. 82), although most measure just a few centimetres. The external surfaces of their shells can be smooth or marked by radiating ribs, concentric ridges and growth rings, the latter often produced annually. A specimen of *Arctica islandica* was shown to be 507 years old by counting the growth rings, making it the longest-lived animal known. Visible on the interiors of modern shells and well-preserved fossils are one or a pair of shallow depressions where the adductor muscles that close the shell are attached. It is also possible to observe structures along the hingeline where the two valves articulate. Ridges in each valve, known confusingly as teeth but having no role in feeding, fit into corresponding sockets of the opposite valve. Also present is a groove containing an elastic ligament serving to open the shell when the adductor muscles are relaxed. The soft body of the animal is cloaked in a thin membrane known as a mantle, which lines the inner surface of the shell. Three folds are present near the outer edge of the mantle: the inner fold is muscular and controls the flow of water in and out of the animal; the middle fold has a sensory function; and the outer fold is involved in shell formation. The position of the inner fold is marked on the shell interior by the pallial line. An indentation in the pallial line called a pallial sinus is present in bivalves possessing siphons, tubular structures through which water is inhaled and exhaled for feeding and respiration, forming a conduit to the surface in bivalves that inhabit burrows. A deep pallial sinus is a useful indication in fossil bivalves that the animal had a long siphon and a burrowing lifestyle.

The preservation of fossil bivalves varies enormously. To a large extent this is due to differences in the mineralogy of the shells belonging to different species, but it also reflects mode of life. The oldest bivalves dating from the Cambrian period had shells made entirely of aragonite. Like many younger Palaeozoic and Mesozoic fossil bivalves, shells composed of this more soluble form of calcium carbonate tend to be preserved as moulds following dissolution of the shell. Occasionally, however, the aragonite shell is replaced by crystalline calcite. Bivalves using calcite to make their shells are far better represented in the fossil record. Although calcite is first found in some Early Palaeozoic bivalves, it was not until the Mesozoic that this biomineral

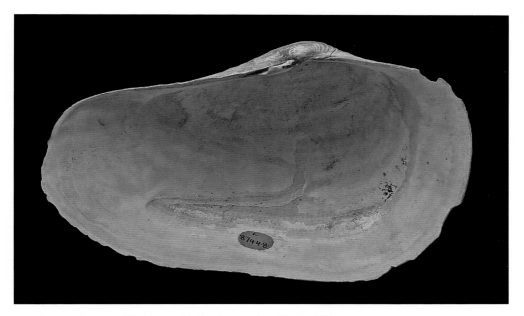

ABOVE Interior of a valve of the Miocene bivalve *Panopea* from Virginia, USA. Visible above the yellow label in this 12 cm (4¾ in) long fossil is an inflection in the pallial line called the pallial sinus, providing evidence that the animal had a long siphon and was a deep burrower.

became common, appearing in numerous groups of bivalves. As a rule, calcite shells occur in bivalves that live on the sediment surface, such as oysters and scallops, whereas shells of burrowing bivalves, such as cockles, are more often aragonitic and typically survive in older rocks as moulds.

While fossil bivalves are present in Palaeozoic rocks, they are greatly outnumbered by brachiopods, which also possess two valves (p. 94). The origins of most major groups of bivalves can be traced back to this era, and many of the lifestyles known among more recent bivalves first evolved in the Palaeozoic. These lifestyles include deposit feeding by processing sediment to extract food, and suspension feeding by filtering small particles of food from the seawater. Some Palaeozoic bivalves were epifaunal, living on the sediment surface; others were infaunal, excavating burrows in the sediment from which they extended siphons into the water above to feed and respire. A few even had the ability to bore holes into hard rocks or wood for protection (see Chapter 13). The Ordovician bivalve *Modiolopsis* lived semi-infaunally, part of the shell being buried in the sediment for anchorage and the remainder being exposed. Because this genus had a thin shell of aragonite, it is frequently preserved as an internal mould (steinkern).

ABOVE The 'devil's toenail' of folklore is the Early Jurassic oyster *Gryphaea arcuata* in which the left valve is strongly curved and the right valve small and flat. This example is 7 cm (2¾ in) long.

ABOVE RIGHT Steinkern of the Ordovician bivalve *Modiolopsis modiolaris* from Ontario, Canada.

The plethora of bivalves found in Mesozoic and Cenozoic rocks makes these molluscs among the commonest of all fossils. Topping the table of abundance are oysters. Fossil oysters range from roughly a centimetre (½ in) to, exceptionally, a metre (39 in) in length. Their shells are made predominantly of calcite and may be very thick, two traits promoting their survival as fossils. The valves of oyster shells usually differ in size and shape. A more convex left valve is positioned lowermost in oysters that live reclined on the seabed, whereas in oysters that cement themselves to hard surfaces, the left valve conforms to the shape of the substrate. *Gryphaea* is a fossil oyster of the reclining kind with a special cultural and scientific interest. Dubbed the 'devil's toenail' because of the talon-like shape of the left valve, powdered *Gryphaea* shells were once believed to cure arthritis. *Gryphaea* was the subject of pioneering research on evolution undertaken by British geologist Arthur Trueman (1894–1956) a hundred years ago. Trueman identified a succession of different species of *Gryphaea* in Early Jurassic rocks covering an interval of about 15 million years. The early species were small, narrow and thick with sharply curved left valves. These were succeeded by species with larger, broader and thinner shells with almost flat left valves. This evolutionary trend has been interpreted in terms of changing strategies for maintaining a stable position on the seabed, from a 'dead-weight strategy' in the thick-shelled early species, to a 'snowshoe strategy' in the broad-shelled later species, with their larger surface area resting on the sediment.

Another important group of bivalves that diversified after the Palaeozoic is the scallop family, Pectinidae. Living pectinids spend most of their lives resting impassively on the seafloor, although many species can swim in short bursts to escape from predators, the valves flapping open and closed to force water out from between them and propelling the animal backwards. The single large adductor muscle used to close the valves in swimming and at other times is the part of the scallop devoured by lovers of seafood. Pectinids have fan-shaped, calcitic valves covered by radiating, fluted ribs. Small triangular outgrowths known as ears are present on one or both sides of the 'umbo' at the pointed end of the shell. Quite often, one of the valves is convex and the other flat, the latter oriented uppermost in living animals resting on the sediment surface. The state fossil of Virginia, USA is a Miocene–Pliocene pectinid, *Chesapecten jeffersonius*, the species name commemorating Thomas Jefferson, a founding father of the USA.

Two remarkable extinct groups of bivalves epitomize rocks of Cretaceous age. Both groups have bimineralic shells with an inner layer of aragonite and an outer, typically thicker, layer of calcite. Each contains some gargantuan species whose huge size has been attributed to the presence of microscopic symbionts. The first is the Inoceramidae. Fragments of inoceramid shells, along with very occasional complete

BELOW Exterior and interior views of the scallop *Chesapecten jeffersonius*. Measuring 6.5 cm (2½ in) wide, this example was collected from the Miocene rocks of Virginia, USA where it is the official state fossil.

RIGHT This specimen of the bivalve *Inoceramus pictus* from the Cretaceous Chalk of Guildford in Surrey, England has importance because it is the type specimen of the species. Unusually, the 11 cm (4¼ in) long shell preserves colour banding.

shells, abound in the Late Cretaceous Chalk Formation. Inoceramid shells often have concentric folds on their outer surfaces. The shell of prismatic calcite makes inoceramids readily recognizable, the shells habitually breaking into small needle-like crystals that can be found scattered throughout the Chalk. A true giant among inoceramids is *Platyceramus platinus*, which reaches up 3 m (10 ft) long in the Cretaceous rocks of Colorado. This species is interpreted to have rested recumbent on the seafloor in the darkness at 200–350 m (656–1,150 ft) water depth where oxygen levels were low. Although this may not seem at all conducive to achieving such a huge size, it has been hypothesized that *Platyceramus platinus* harboured chemosymbiotic bacteria, which acted as an energy source and allowed the host bivalve to grow so large. This theory has received support from studies showing a similar isotopic composition in inoceramid shells to some modern bivalves known to harbour chemosymbionts.

The second bivalve group containing species with huge shells is the order Hippuritida, better known as rudists. Evolving in the Late Jurassic and becoming extinct near the end of the Cretaceous period, rudists inhabited the shallow warm waters on the edges of the Tethys Ocean which extended east–west across what is now central America to southern Europe and the Middle East. Sections of rudists can frequently be observed in building stones used in these regions. An enormous variety of shell shapes occurs among rudists. Some have two openly coiled valves, the shell as a whole resembling a pair of cattle horns. More typical are rudists with a large conical valve capped by a smaller, lid-like valve, at first sight easily mistaken for a solitary coral (p. 116). Some exceed a metre in length and have extremely thick shells, leaving remarkably little space for the soft parts of the animals between the valves. Along with the fact that rudists often grew in clusters on the seafloor to form reef-like structures, their large and thick shells have led palaeontologists to hypothesize

ABOVE Limestones full of rudist bivalves are commonly employed as building stones, particularly in southern Europe. Sectioned rudist shells can be seen here in the stone slabs paving the concourse of Porto railway station in Portugal.

ABOVE RIGHT Many rudist bivalves resemble horns, as in this example of *Hippurites* from the Cretaceous rocks of southern France. The thick-shelled fossil is about 20 cm (7¾ in) long.

that algal photosymbionts, analogous to the zooxanthellae found in the giant clam *Tridacna* today, were present in the mantle tissues of these bivalves.

Trigoniids are another group of bivalves common in Mesozoic rocks. Unlike inoceramids and rudists, this family has survived to the present day, although it is represented by only one genus, *Neotrigonia*, a burrowing bivalve living in the seas around Australia. The asymmetrical valves of trigoniids are divided into two parts, a larger part covered in strong ribs often adorned by tubercles, and a smaller 'escutcheon' that is more delicately ornamented. The thick shell with its complex hingeteeth is composed entirely of aragonite, which accounts for the fact that fossil trigoniids are often preserved as moulds following dissolution of the shell material. Such is the case in the Late Jurassic Portland Stone of southern England, most spectacularly in a rock type called 'Portland Roach' used as a decorative building stone. External moulds of the trigoniids *Laevitrigonia* and *Myophorella* appear as cavities on the surface of

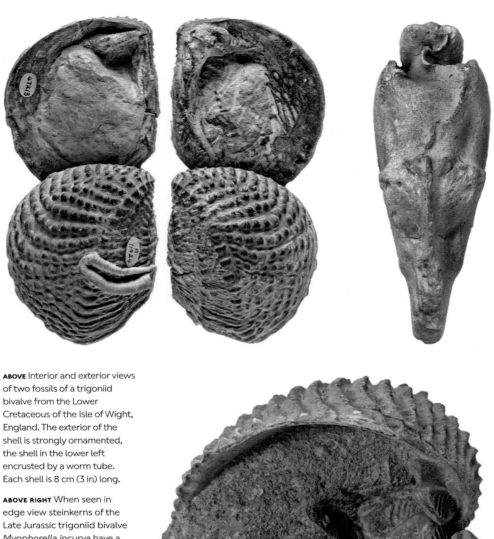

ABOVE Interior and exterior views of two fossils of a trigoniid bivalve from the Lower Cretaceous of the Isle of Wight, England. The exterior of the shell is strongly ornamented, the shell in the lower left encrusted by a worm tube. Each shell is 8 cm (3 in) long.

ABOVE RIGHT When seen in edge view steinkerns of the Late Jurassic trigoniid bivalve *Myophorella incurva* have a peculiar appearance which led to their folklore name 'Osses 'Eds. This fossil from the Isle of Portland in Dorset, England is 8 cm (3 in) high.

RIGHT In this example of *Trigonia crenulata* from the mid Cretaceous of Le Mans, France, the two valves are open, revealing the hinge line. The fossil is 8.5 (3¼ in) cm across.

Portland Roach, preserving the ribs and other features of the shell exterior in negative relief. Hardening of lime mud that infilled the space between the two valves during burial produced internal moulds (steinkerns) so unusual in appearance that their identity puzzled early naturalists.

A burrowing mode of life and an aragonitic shell characterize two other groups of bivalves that are usually preserved as internal moulds in Mesozoic rocks. The anomalodesmatan *Pholadomya* first appeared in the Late Triassic, becoming abundant and diverse in the Jurassic and Cretaceous before subsequently dwindling. The last surviving species is the extremely rare *Pholadomya candida* from the Caribbean Sea. Generally believed to be a deep burrower, *Pholadomya* possessed a long siphon that passed through a space or gape between the valves and extended into the water above for suspension feeding and respiration. The shell of *Pholadomya*, which is extremely thin and covered by tiny spikes, is seldom fossilized. However, steinkerns of *Pholadomya* and the related *Pleuromya* are common and can be found among stones brought up by ploughs in fields underlain by Jurassic limestones in England.

The bivalves so far described were marine animals. In contrast, bivalves belonging to the suborder Unionida inhabit lakes and rivers and include freshwater pearl mussels. Unionids have been found as fossils in non-marine deposits as far back as the Permian. Fossil unionids commonly attributed to the extant genus *Unio* are especially abundant in some freshwater sedimentary rock.

BELOW Retaining its brilliant lustre, this shell of the bivalve *Pholadomya virgulosa* comes from the Early Eocene London Clay of Portsmouth in Hampshire, England. The fossil is 8 cm (3 in) across.

BELOW RIGHT Freshwater bivalves, commonly identified as *Unio,* cover a bedding plane in this 20 cm (7¾ in) wide piece of Early Cretaceous rock from the Weald of southeast England.

Shell beds rich in aragonitic molluscs become common in the Cenozoic and often contain pristine shells belonging to large numbers of species. For example, the Pliocene and Pleistocene rocks of southern Florida, USA are a better source for shell collectors of large, beautifully preserved molluscs than can be found on the modern beaches.

The thick, chalky white aragonitic shells of *Venericor* are common in the Eocene rocks of southern England and elsewhere in Europe and North America. Shells of this bivalve have a curved umbo, broad, flat ribs on the exterior, and a prominent pallial line extending between two adductor muscle scars. Some species were shallow burrowers, the ends of the shells poking above the sediment surface. *Venericor* occasionally fell victim to predatory octopus, as shown by the presence in their shells of the diagnostic drillholes made by these cephalopods.

Whereas *Venericor* is extinct, another shallow-burrowing genus – *Glycymeris* – which is common as a Cenozoic fossil, persists in shallow seas at the present day. The extant species *Glycymeris glycymeris*, known as the dog cockle, usually inhabits gravel bottoms and has almost bilaterally symmetrical valves with numerous small hingeteeth. Individuals of this species are remarkably long-lived, some reaching an age of 200 years.

RIGHT Two dark-coloured muscle scars are clearly visible in the interior of a shell of the bivalve *Venericor*. Measuring 10.5 cm (4 in) across, this fossil was collected in Eocene rocks of the Bracklesham Group during excavations for a dock extension in Southampton, Hampshire, England.

LEFT The hingeteeth are clearly seen in this view of the interior of a *Glycymeris* shell from Plio-Pleistocene rocks of eastern England. The shell is 4.7 cm (1¾ in) in height.

Glycymeris is one of many molluscs abounding in the Plio-Pleistocene Red Crag, which outcrops over parts of Essex and Suffolk in eastern England. Like the other bivalves in this formation, shells of *Glycymeris* are invariably disarticulated, and were abraded during transportation before being buried beneath advancing submarine dunes.

Gastropod molluscs

The European Garden Snail *Cornu aspersum* is just one of approximately 80,000 species of gastropods alive today. Gastropods are a remarkably successful class of molluscs, not only in terms of their high diversity, but also in the range of ecological niches they have evolved to occupy. Different species of gastropods can be found living on land, in freshwater lakes and rivers, and in the oceans from the intertidal zone down to a depth of 10 km (6¼ miles). Some graze on algae and other plants, whereas others are carnivores or feed on food particles suspended in the water or contained in the sediment. Most have calcareous shells, although shells have been lost over evolutionary time in terrestrial slugs and marine sea slugs (nudibranchs). Not surprisingly, the shell-less groups are very seldom fossilized and, unfortunately, the overwhelmingly aragonitic composition of gastropod shells has depleted the fossil record of the class, especially in older rocks.

Unlike bivalves, gastropod shells consist of just one valve. This is coiled into a conical spiral shape known as a helicospiral. Gastropod shells vary in the degree to which the helicospiral is three-dimensional, from almost flat-coiled, planispiral species to elongate, high-spired species, while in limpets the helicospiral coiling is barely perceptible, the shell being cap-shaped. Another source of variation among gastropods is the ornamentation of the shell surface. This can take the form of ridges oriented either parallel (cords) or at right angles (ribs) to the coiling direction, as well as tubercles and spines. The shape of the aperture from which the foot of the animal emerges also varies, from almost circular to crescent- or slit-shaped. Taken together, these variations allow a large number of living and fossil gastropods to be distinguished using the shape of the shell alone.

RIGHT *Symmetrocapulus tessoni*, a limpet shell 4 cm (1½ in) in size from the Middle Jurassic rocks of Minchinhampton in Gloucestershire, England.

BELOW RIGHT Fossils of the gastropod *Oriostoma discors* are common in Silurian rocks of the Welsh Borderland. This example from Coalbrookdale in Shropshire, England measures 6 cm (2¼ in) in diameter.

As with bivalves, fossils of gastropods increase in diversity and abundance from Palaeozoic through to Mesozoic and Cenozoic rocks. The oldest known gastropods from the Early Cambrian are small, just a few millimetres in size, often limpet-shaped and have been found particularly in marine deposits where fossil shells have been replaced by phosphate minerals. Larger gastropods become more abundant in Ordovician and Silurian rocks. *Oriostoma discors* (formerly known as *Poleumita discors*) is a euomphalid gastropod commonly found in the Silurian Wenlock Limestone of the Welsh Borderlands. Occasional specimens are found with an intact operculum, a protective lid that was used to close the aperture when the foot of the gastropod was retracted. Fossilization of the operculum in *Oriostoma* occurred because it was calcified, a condition found in a minority of fossil and living gastropods and associated with species that boxed themselves in rather than attempting to flee from predators.

The planispiral shells of the Cambrian–Triassic genus *Bellerophon* are often collected from the Lower Carboniferous limestone of the British Isles. Named after the Greek hero, famous for slaying the monster Chimera, this and closely related genera were once universally regarded as gastropods, but it is now thought that at least some may have belonged to the Monoplacophora, a class of primitive shell-bearing molluscs lacking the diagnostic 180° twisting (torsion) of the body relative to the shell during larval development seen in gastropods. *Bellerophon* has a globose shell on first sight easily mistaken for a cephalopod but without septa to subdivide the shell chamber.

Platyceratids are a family of Palaeozoic gastropods of interest because of their persistent association with crinoids, a class of suspension-feeding echinoderms (p. 129). They are often found attached to fossil crinoids with which they had a symbiotic relationship. It was once believed that platyceratids fed on the faeces of the

LEFT Coiled in one plane, this Early Carboniferous fossil of *Bellerophon tangentialis* was found near Clitheroe in Lancashire, England more than 150 years ago. The shell measures 6 cm (2¼ in) across.

RIGHT Vertically sectioned fossil of the Early Jurassic gastropod *Pleurotomaria* almost 5 cm (2 in) in height. The chambers of the shell are filled by a mixture of hardened mud and crystals of calcite that grew after burial.

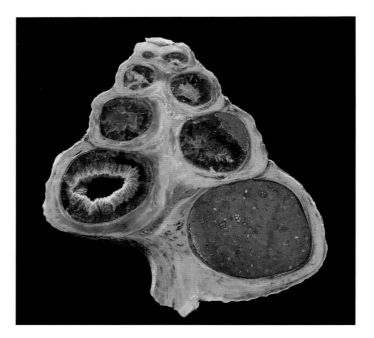

BELOW RIGHT Steinkern of the Late Jurassic gastropod *Aptyxiella portlandica* from Dorset, England. Fossils like this result from lime mud filling and hardening inside the shell, followed by dissolution of the shell itself.

crinoid, but they may alternatively have stolen food from the crinoid (kleptoparasitism) with some species known to have bored through the skeleton to access and consume parts of the host crinoid.

Mesozoic gastropods have shells of more varied shape, particularly in the Cretaceous when predatory lifestyles were widely adopted. Two genera found in the British Jurassic are useful in illustrating contrasts in shell shape and preservation. *Pleurotomaria* is a so-called slit shell, named for the presence of a fissure extending backwards from the shell aperture functioning as an outlet for respiratory currents. Slit shells are an ancient group, first known from the Late Cambrian. Although living slit shells inhabit only deep-sea environments, in the geological past they colonized shallow waters too. Fossils of *Pleurotomaria* have the original shell preserved because it is one of a small proportion of genera with a shell made of calcite; other calcitic gastropods include most limpets, a few periwinkles (*Littorina*) and wentletraps. *Aptyxiella*, on the other hand, had an aragonitic shell, which almost invariably dissolved during fossilization, leaving behind an exterior mould and, if the shell cavity was infilled by sediment, an internal mould or steinkern. This is the case in the 'Portland Roach' where this gastropod is found together with the trigoniid bivalves described above. Because *Aptyxiella* is a high-spired gastropod, the internal moulds have the shape of a screw, hence the English folklore name 'screwstone' for these fossils.

Regardless of the increased probability of aragonite shells surviving in younger deposits, gastropods became more prolific during the Cenozoic. The largest known gastropod ever to have existed is *Campanile giganteum* from the Eocene, with high-spired shells reaching up to 90 cm (3 ft) in length. Judging by the sole living species of this genus, the Eocene species was a gentle giant that fed on algae. In contrast, the far smaller *Natica* is not only a carnivore but also often a cannibal. These gastropods drill distinctive holes (see Chapter 13) into the shells of other molluscs, including those of the same species, to gain access to the meaty parts within. The globular shells of *Natica* are often found in rocks ranging from the Paleocene to the present day, especially in sands and silts into which it burrowed to locate its prey.

Almost all gastropod shells are right-handed (dextral) – the shells coiling in a clockwise direction as they grow, and the aperture lying on the right side when the shell is viewed from the front with the apex pointing upwards. Left-handed (sinistral) gastropods have evolved in several groups through time, but all quickly became extinct. One example is *Neptunea angulata*, which is extremely common in the Plio-Pleistocene Red Crag Formation. Although the sinistral *Neptunea angulata* is extinct, the genus is still represented today by dextral species such as the Red Whelk *Neptunea antiqua*. A persuasive explanation has yet to be offered for the evolutionary failure of sinistral species.

It has been estimated that more than 30 groups of gastropods independently made the transition from the sea into freshwater environments. There are many fewer fossils of freshwater gastropods for two main reasons: they had thin shells, which were less easily preserved, and some of the environments they inhabit, such as streams, are seldom represented in the geological record. Nevertheless, there is evidence that freshwater gastropods had evolved by Carboniferous times at the very latest. Freshwater gastropods belonging to several extant groups appeared in the Mesozoic. Limestones packed with the shells of *Viviparus* from Early Cretaceous freshwater lakes in southern England are known as 'Purbeck Marble' and 'Wealden Marble'. Striking when polished, these sedimentary rocks are used decoratively on table-tops and in architectural features such as pillars and flooring.

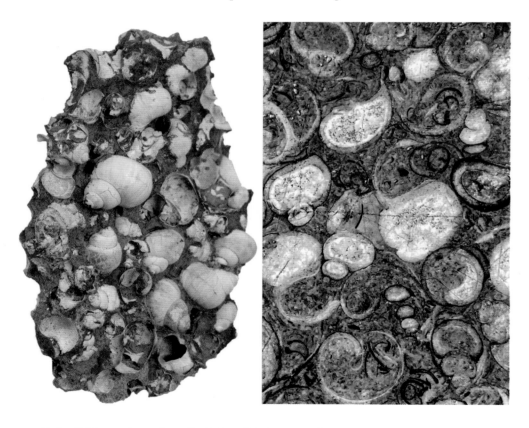

ABOVE Shells of *Viviparous lentus* abound in this 9 cm (3½ in) wide rock from the Eocene–Oligocene Hempstead Beds, England.

ABOVE RIGHT Detail of a tabletop made from polished Wealden Marble, an Early Cretaceous limestone packed with shells of the gastropod *Viviparus*.

Other molluscs

It would be remiss not to mention three other classes of molluscs – Polyplacophora, Scaphopoda and Rostroconchia – that can be found as fossils, even though they are much rarer than bivalves and gastropods. Otherwise known as chitons, polyplacophorans have eight articulated plates of aragonite forming a protective cover over the upper surface of the body. They live on hard substrates, moving slowly around and, like a limpet, use teeth (radulae) reinforced with the hard mineral magnetite to scrape off algae and attached animal prey. After death, the plates disarticulate and are usually found scattered in fossil specimens but can be reassembled. The fossil record of polyplacophorans can be traced back to the Cambrian but they are under-represented as fossils, not only because of the aragonitic composition of the plates but also because of living in erosive, rocky shore environments.

Scaphopods or tusk shells look like broken stems of slightly curved clay pipes, tubular and open at both ends and typically a few centimetres long. Some have a

TOP Assembled plates of the polyplacophoran *Gryphochiton priscus* from the Early Carboniferous of Tournai in Belgium mounted on a card. The reconstruction measures 6.5 cm (2½ in) long.

ABOVE Measuring 8.7 cm (3½ in) in length, this is the fossil of a scaphopod shell from the Pliocene of the Italian island of Sicily.

RIGHT A small rostroconch mollusc belonging to the genus *Conocardium* from the Carboniferous of Belgium, measuring 14 mm (½ in) across.

smooth shell, but in others the shell is ribbed. Scaphopods burrow using a foot that extends from the opening at the broader end of the shell. The foot bears small tentacles, which remove food items including foraminifera (see Chapter 12) from the sediment. The shells of scaphopods have been found as fossils back to the Carboniferous but their aragonitic composition biases against their fossilization, and some supposed scaphopod fossils have turned out to be serpulid worm tubes (see p. 124).

First recognized as a distinct class in 1972, rostroconchs are an extinct, exclusively Palaeozoic group of molluscs. They have a single, taco-shaped shell resembling a bivalve in which the two valves are fused along the hingeline. Rostroconchs were probably suspension feeders inhabiting shallow burrows, with a flattened tubular extension at one end of the shell bringing seawater laden with plankton to the animal for processing.

Brachiopods

Today, almost all invertebrates with a pair of articulated calcareous shells are bivalve molluscs. Turn the clock back 250 million years or more into the Palaeozoic era and animals belonging to the phylum Brachiopoda had the monopoly on the two-valved design and can be found in great abundance in rocks of this age. Sometimes known as lampshells because of their superficial resemblance to Roman oil lamps, brachiopods are still present in modern seas but are represented by about 350 species only, a paltry number compared with the estimated 12,000 fossil species. Most people will never encounter a living brachiopod as they are rare and, except for one genus (*Lingula*), are not eaten by humans. They are more common in the southern hemisphere, but even here they are seldom conspicuous.

One of the first lessons learned by students of palaeontology is how to distinguish brachiopods from bivalve molluscs. In most cases, this can be done using differences in their symmetry: the two valves in a bivalve shell are mirror images of each other, meaning that a plane of bilateral symmetry passes between the valves; whereas in brachiopods the two valves are dissimilar, and the plane of bilateral symmetry runs vertically through the centres of the valves. The larger of the two valves in a brachiopod is called the ventral valve and has a beak-like extension (umbo) overhanging the top of the smaller dorsal valve. The umbo contains a hole through which emerges the pedicle, a muscular organ used to anchor the living animal to a substrate on the seafloor. In some brachiopod species, referred to as 'ambitopic', the pedicle is lost during development and the hole is closed, the shell either coming to lie freely on the sediment surface or, in a few species, being affixed to a hard substrate via a cemented ventral valve.

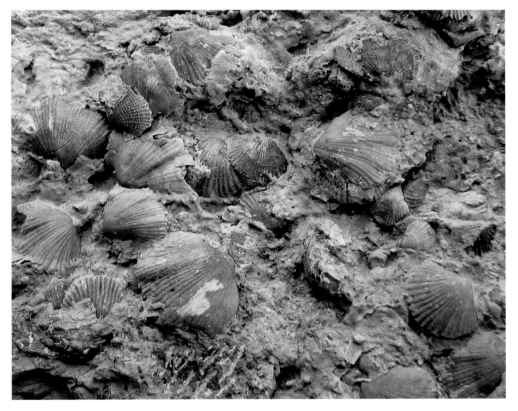

ABOVE Field photograph of brachiopods in limestone belonging to the Late Ordovician Bromide Formation of Oklahoma, USA. Individual shells are 2–3 cm (¾–1 in) in size.

There are other important differences between brachiopods and bivalve molluscs. Brachiopods use muscles both to open and close their shell. In bivalves an elastic ligament is responsible for opening the shell and muscles are employed only during closure. This has implications for the preservation of brachiopod and bivalve fossils because decay of the muscles following death routinely allows the elastic ligament to spring the valves open to a 'butterfly' posture in bivalves, whereas in brachiopods the valves often remain together – closed and fully articulated. As a result, fossil brachiopod shells commonly pop neatly out of the rock when it is eroded or struck with a hammer.

Brachiopods failed to evolve the great variety of lifestyles seen in bivalve molluscs. All brachiopods are surface suspension feeders employing a body part called a 'lophophore' to capture plankton; there are no deposit feeders processing sediment for food. Few brachiopods burrow and none can bore deeply into hard substrates. There are no swimming brachiopods, although some recent species have been observed to use the pedicle as a kind of pogo stick to move around. Brachiopods have also failed to colonize freshwater environments. Nevertheless, brachiopods did once occupy many of the ecological niches that are now the domain of bivalves. The switch from brachiopod supremacy in the Palaeozoic to bivalve dominance in the Mesozoic and Cenozoic has elicited various explanations. Some palaeontologists contend that competition from bivalves brought about the demise of brachiopods. Others have dismissed this idea, characterizing the two groups as 'ships that pass in the night', favouring factors other than competition, such as the devastating effects of the end-Permian mass extinction on brachiopods.

Modern classifications recognize three subphyla of brachiopods: Linguliformea, Craniiformea and Rhynchonelliformea. Linguliform brachiopods have shells made of calcium phosphate and collagen fibres, whereas craniiform and rhynchonelliform brachiopods have calcium carbonate shells, invariably calcitic. All three subphyla are extant, linguliforms and rhynchonelliforms first appearing in the Cambrian and craniiforms slightly later in the Ordovician.

The best known linguliform brachiopod is *Lingula*. The phosphatic shells of *Lingula* resemble slender fingernails in shape and are often dark brown to black in colour. Although typically found in brackish water environments, these brachiopods are widely distributed in tropical seas today. *Lingula* excavates a vertical burrow in the mud to a depth of 10 cm (4 in) or so, the pedicle anchoring the shell to the bottom of the burrow and pulling it away from the surface when protection is required. *Lingula* has a geological range exceeding 500 million years from the Cambrian to the present day and has been dubbed a 'living fossil' because its morphology seems to have scarcely changed through this enormous expanse of time.

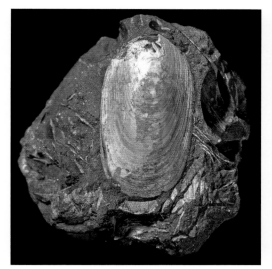 

ABOVE Collected from Jurassic rocks in Yorkshire, England this is the brachiopod *Lingula beani*. Growth lines on the shell surface are clearly visible.

ABOVE RIGHT Six individuals of living *Lingula* excavated from the mud. Note the long pedicle that anchors the animal in the sediment.

Craniiforms were once grouped with linguliforms as inarticulate brachiopods, because the two mineralized valves are not articulated to each other. They are, instead, held together only by muscles. Deep pits in fossils of *Crania* and its close relatives show where these muscles were attached to the internal surfaces of the calcareous valves. Paired adductor muscle pits on either side of a ridge can resemble eyes and a nose, giving the shell the appearance of a cartoon face. Fossils of craniiform brachiopods are not uncommon in the Late Cretaceous Chalk encrusting other shells. They are invariably disarticulated with only the cemented ventral valve preserved in situ; the low, limpet-shaped dorsal valves are to be found loose in the sediment.

Rhynchonelliforms are by far the commonest and most diverse fossil brachiopods, encompassing most of what were once known as articulate brachiopods. Only a few representatives are highlighted here. The orders Orthida and Strophomenida first appeared in the Cambrian and diversified as part of the 'Great Ordovician Biodiversification Event' to become the dominant brachiopods of the Ordovician. Orthids characteristically have shells with ribs radiating from the umbo and an almost straight hingeline along which the two gently convex valves articulated. Strophomenids differ in typically having a flat or concave dorsal valve sitting like a lid within the convex, and appreciably larger, ventral valve. Most species of both orders were ambitopic, discarding the pedicle during their development and becoming recumbent on soft substrates or, in a few cases, cemented

RIGHT The interior surface of shells of the craniiform brachiopod *Ancistrocrania antigua* often resemble cartoon human faces. In Sweden such fossils were once known as 'Brattingsborg pennies'. This specimen from the Late Cretaceous Chalk of Belgium measures about 1 cm (½ in) across.

by the ventral valve to hard surfaces. The life orientation of a strophomenid called *Rafinesquina*, which is present in vast numbers in Late Ordovician rocks of the US mid-west, has been hotly debated. Some palaeontologists believe *Rafinesquina* to have rested on the dorsal valve with the convex ventral valve uppermost, a view based on the common presence of encrusting bryozoans (p. 116) etc. on the ventral valve and the greater hydrodynamical stability of this orientation in currents. Others consider that *Rafinesquina* rested on the ventral valve with the concave dorsal valve uppermost, thus elevating the gape of the shell above the sediment to make mud less likely to enter the lophophore during feeding.

Brachiopods of the rhynchonelliform order Spiriferida are especially abundant in the Devonian, although their total range is Ordovician to Jurassic. Their elegant shells are typically broad, resembling a pair of wings, and marked by conspicuous growth lines and radial ribs, along with a median fold or sulcus. Helicospirally coiled internal ribbons within each of the 'wings' acted as skeletal supports for a spiral lophophore. Spiriferids are believed to have fed by creating an incurrent of water at the median sulcus, which then flowed through the spiral lophophore for extraction of plankton before being vented laterally at the left and right sides of the shell. Beautifully preserved spiriferids abound in the Middle Devonian rocks of northeastern USA and Ontario, Canada. Devonian slates quarried at Delabole in Cornwall, UK contain spiriferids of far inferior preservation, many stretched during metamorphism of the rock, but notable in retaining the double-winged form, which led to their folklore name 'Delabole Butterfly'.

Productids are an important brachiopod order that came to prominence in the Carboniferous and Permian. They usually have thick shells comprising a large,

ABOVE Two fossils of the Late Ordovician strophomenid brachiopod *Rafinesquina* from the Cincinnati region of the USA, each about 4.5 cm (1¾ in) wide. The shell on the left shows the exterior of the dorsal valve, that on the right the interior.

LEFT The interior of this ventral valve of the orthid brachiopod *Multicostella sulcata* was encrusted after death by a branching colony of the bryozoan *Zigzagopora wigleyensis*. From the Late Ordovician Bromide Formation of Oklahoma, USA, the shell is about 1.5 cm (½ in) across.

BELOW LEFT Spiriferid brachiopod *Mucrospirifer* from Devonian rocks of Ohio, USA. This view of a dorsal valve measuring 3.5 cm (1¼ in) across shows the characteristic median fold.

convex ventral valve and a smaller more-or-less flat dorsal valve. Hollow spines are often present on the shell surface, some probably having a defensive function while others were able to grow around and clasp objects such as crinoid stems to improve the stability of the brachiopod. Compared with spiriferid brachiopods, it has been suggested that productids were able to flourish in environments with a low supply of planktonic food. The Carboniferous productid *Gigantoproductus* is the largest known brachiopod, with a shell up to 30 cm (1 ft) in width but a surprisingly narrow space between the two valves in which the lophophore and other soft tissues would have been accommodated. Shell beds and lens-shaped accumulations of *Gigantoproductus* attest to the success of this titan among brachiopods.

Despite their reduced importance after the Palaeozoic era, brachiopods are still common fossils in many Mesozoic and Cenozoic marine rocks. Two orders dominate: Rhynchonellida and Terebratulida. Both have short, curved hingelines contrasting with the long, straight hingelines of the Palaeozoic orders described above, and tend to have shells with an ovoidal to rounded triangular outline shape. Rhynchonellids usually have strongly pleated valves. The pleats are angular and create a zig-zag opening around the edge of the shell where the two valves meet. Not only do the pleats strengthen the shell but the zig-zag opening increases the area through which plankton-laden water can flow while retaining a narrow gape preventing the entry of large, unwanted particles.

Terebratulids generally have smooth shells which, if examined closely, can be seen to contain minute pores called punctae, useful in distinguishing between shell fragments of these brachiopods and molluscs. They are especially common in the Middle Jurassic limestones outcropping from the Cotswolds to the Dorset coast in southern England. As a child William Smith, who pioneered the use of fossils in

RIGHT Ventral valve of a 16 cm (6¼ in) wide specimen of the productid brachiopod *Gigantoproductus* from the Early Carboniferous.

ABOVE Views of the ventral (top) and dorsal (bottom) valves of the rhynchonellid brachiopod *Cyclothyris difformis* from Cretaceous rocks of Devon, England. The shell is about 3 cm (1 in) wide.

ABOVE RIGHT This large terebratulid brachiopod *Pliothyrina grandis* from the Pliocene Coralline Crag of Orford in Suffolk, England measures 10 cm (4 in) in height. The opening through which the pedicle once emerged is visible at the top of the fossil.

stratigraphical correlation (p. 21), would collect these brachiopods from the fields around his home in Churchill, Oxfordshire, UK. Known to the locals as 'pundibs', the children here used them in their games. The largest fossil terebratulid found in Britain is also one of the youngest, *Pliothyrina maxima* from the Pliocene Coralline Crag of Suffolk, in which the shell reaches up to 10 cm (4 in) in length. Today, some four million years later, there are no brachiopods approaching this size in the seas around Britain.

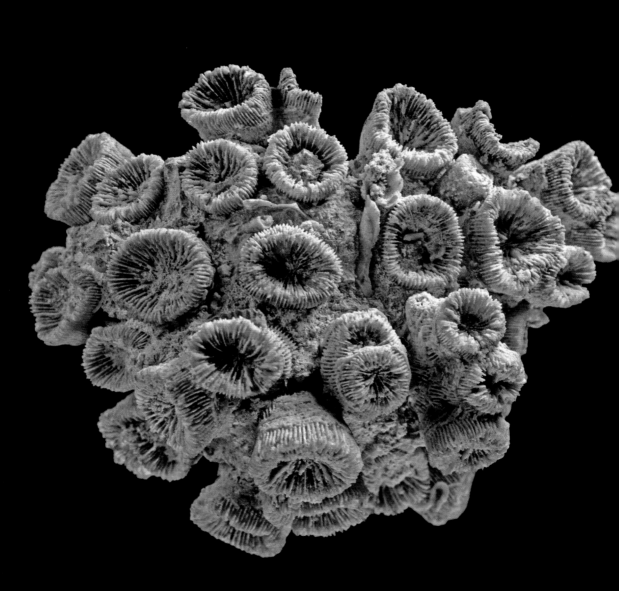

6 Invertebrates of all shapes and sizes

The oceans abound with invertebrates carving out a humble living by feeding on small particles of food suspended in the water around them. Because water currents carry the food particles towards these animals, there is no necessity for them to be mobile – they can live attached to rocks or other hard surfaces, or root themselves into the sediment, remaining in one place throughout their lifespans. Like diners in a sushi restaurant where a conveyor belt delivers plates of food to be selected at will, these animals only require a way of capturing passing morsels of plankton. The ability to feed while remaining essentially static brings with it the massive advantage of not needing to spend energy chasing around after prey. On the other hand, immobility makes these sessile animals sitting targets for aspiring predators. While some rely on chemical defences to avoid falling victim to predators, many enclose their soft parts in hard exoskeletons which can survive long after death, which is why these animals are among the best represented in the fossil record.

Another trait common to many of the invertebrates described in this chapter is coloniality. Colonial animals are aggregates of genetically identical modules, known as zooids, originating from a single founding individual. Colonies represent a kind of society in which the welfare of individual zooids is less important than that of the colony as a whole. As with human societies, colonies can exhibit division of labour, some zooids being specialized to perform specific life functions such as feeding, defence or reproduction. Colonies display a phenomenon known as 'partial mortality' whereby zooids may die because of old age, or by being consumed by a predator, while the colony continues to live. Immobile suspension-feeding animals are so unlike the dogs, birds, spiders and other animals we encounter on a daily basis, that comprehending their lifestyles is difficult, all the more so when they are colonial.

Several animal phyla common as fossils fall within the remit of the current chapter. The soft-part anatomy of these phyla is distinctive, but their fossilized hard parts can

OPPOSITE Colony of the scleractinian coral *Thecosmilia trichotoma*. Collected from the Late Jurassic rocks at Nattheim in Germany, this fossil measures 11.5 cm (4½ in) across.

be alarmingly similar, causing problems in identification. For instance, some fossils once believed to be corals are now considered to be sponges, while certain corals are routinely misidentified as bryozoans.

Sponges

To most people the word sponge conjures up a soft, squidgy object used in the bath or for washing a car. It is not surprising that the popular perception of sponges is of flimsy animals unlikely to be fossilized; yet many species of sponges have calcareous or siliceous skeletons, which accounts for their common preservation as fossils.

The phylum Porifera – sponges – lies at or very close to the base of the animal tree of life. Exclusively aquatic, sponges differ from all other multicellular animals in lacking cells organized into tissues or organs. The great majority of the 8,600 species living today feed by drawing water through tiny pores in the body, extracting small particles of food, including bacteria, and then expelling the filtered water through larger openings called oscula. An exception are some deep-sea carnivorous sponges equipped with sticky, Velcro-like surfaces to trap small animals. Sponge skeletons generally consist of numerous small needle-like spicules which can be made of calcium carbonate or silica, or of organic fibres composed of collagen. Some sponges have thick basal skeletons of calcium carbonate and do not always have spicules. Although these 'hypercalcified sponges' are rare at the present day, they were extremely common at certain times in the geological past.

RIGHT The archaeocyath *Metaldetes* from the Cambrian of South Australia. Sections of several individuals are visible in this weathered piece of limestone measuring 5 cm (2 in) across.

LEFT Large star-shaped spicules are visible on the surface of this *Astraeospongia* from the Silurian of Tennessee, USA. The fossil measures about 6.5 cm (2½ in) across.

Sponges vary greatly in size and shape. Many grow as crust-like coverings on rocks and shells, others are cup-shaped, stalked, or complexly branched. Even within a single species, the shape can be variable and irregular, making it less useful for identification than in many other animal phyla. Identifying fossil sponges can be extremely challenging! There are four classes of living sponges: (1) Demospongiae, which constitute over 80% of species and usually have siliceous spicules and/or collagen fibres; (2) Hexactinellida, the so-called 'glass sponges', characterized by siliceous spicules with four or six points; (3) Calcarea, which have calcareous spicules; and (4) Homoscleromorpha, a minor group, often without spicules and poorly represented in the fossil record.

Despite the primitive morphology of sponges among animal phyla and extrapolations about their ancient origins from molecular data, fossils that are likely to be sponges have only recently been recognized in Precambrian rocks following the description of an Ediacaran sponge from South China. Fossil spicules point to the presence of all three major sponge classes (demosponges, hexactinellids and calcareans) in the Cambrian. Spectacular examples of completely preserved sponges have been found in the Burgess Shale of British Columbia, Canada and some other Cambrian Lagerstätten. Cambrian rocks also host a short-lived group of animals called archaeocyaths, generally thought to be sponges without spicules. The calcareous archaeocyath skeleton typically comprises two porous cones of calcium carbonate, one inside the other. Assuming archaeocyaths are correctly interpreted as sponges, it is likely that they fed by extracting food particles from water pumped through the pores in the cones, before passing it into the central cavity for expulsion through the open top. Archaeocyaths are particularly interesting because they were responsible for constructing some of the oldest animal reefs known. Younger Palaeozoic rocks contain numerous types of sponges that are difficult to relate to extant groups. Among these is *Astraeospongia*, a saucer-shaped sponge quite common in the Silurian and Devonian. The surface of *Astraeospongia* is covered with thick calcareous spicules in the form of six-rayed stars.

Until quite recently, two groups of predominantly Palaeozoic hypercalcified sponges – chaetetids and stromatoporoids – were thought to be cnidarians. Chaetetids were generally classified as tabulate corals and stromatoporoids as hydrozoans. Chaetetids have skeletons consisting of tightly packed, narrow cylindrical tubes with frequent cross-partitions. They are particularly common in Carboniferous rocks, forming thick, dome-shaped or tabular structures that can be metres in diameter. Stromatoporoids, not to be confused with stromatolites (p. 32), range from Ordovician to Carboniferous and were important reef-building animals during the Silurian and Devonian, often in collaboration with tabulate corals (p. 112). Their skeletons are strongly layered, with successive layers bridged by vertical pillars. Regular hummocks on the surfaces of well-preserved stromatoporoids mark sites where filtered water was vented from the living sponge. Stromatoporoids are typically dome-shaped, with a flat or concave underside,

ABOVE Vertically sectioned specimen of *Chaetetes radians* showing the calcareous tubes of the basal skeleton that were covered by the living tissues of the sponge. This Carboniferous specimen is about 5 cm (2 in) wide and was collected from a locality near Moscow in Russia.

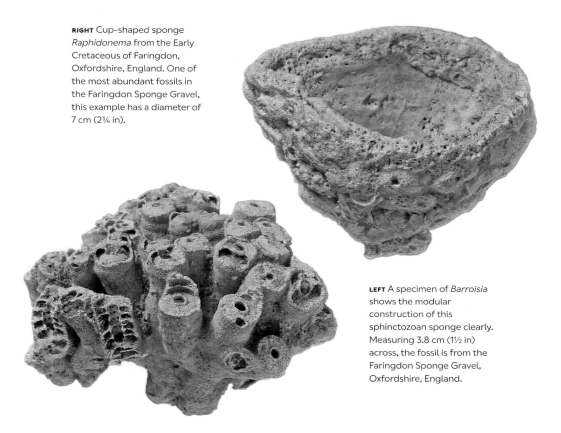

RIGHT Cup-shaped sponge *Raphidonema* from the Early Cretaceous of Faringdon, Oxfordshire, England. One of the most abundant fossils in the Faringdon Sponge Gravel, this example has a diameter of 7 cm (2¾ in).

LEFT A specimen of *Barroisia* shows the modular construction of this sphinctozoan sponge clearly. Measuring 3.8 cm (1½ in) across, the fossil is from the Faringdon Sponge Gravel, Oxfordshire, England.

which contained no living tissues and so frequently became colonized by encrusting or boring animals. One of the best places to see stromatoporoids in the field is on the Swedish island of Gotland. Here, coastal erosion has left spectacular stacks (*rauks*) on the beaches, some of which consist of vertical piles of stromatoporoids.

It can be easier to find fossil sponges in younger rocks. There was a proliferation of species in the Mesozoic with fused or interlocking spicules – and these do not become scattered after death. Fossil sponges of Mesozoic and Cenozoic age look far more like species living in the sea today. Many are lithistids, or rock sponges, an informal grouping of demosponges with rigid, siliceous skeletons. Occasionally, the silica in lithistid skeletons was dissolved soon after burial and replaced by calcium carbonate, as in *Platychonia*, a sponge that constructed the small reefs exposed in the Middle Jurassic rocks of the Calvados coast of France. During the Late Jurassic, larger sponge reefs containing multiple species became widespread across Europe and beyond. Dominated by demosponges but including some calcareans, these reefs are believed to have grown in water exceeding 100 m (328 ft), deeper than contemporaneous coral reefs.

RIGHT *Siphonia tulipa* is a stalked sponge found in Cretaceous rocks. This 14 cm (5½ in) tall specimen was found at Blackdown in Devon, England.

FAR RIGHT Rooted sponge *Craticularia* in a block of Late Cretaceous Chalk. Collected from near Margate in Kent, England.

A good place to find fossil sponges in England is the historic market town of Faringdon, which lies between Oxford and Swindon. The local Faringdon Sponge Gravel is packed with fossils filling a channel in the Early Cretaceous seafloor. Fossils from Faringdon have been exciting naturalists since the seventeenth century when the Keeper of the Ashmolean Museum in Oxford Edward Lhwyd wrote about them. One of the commonest fossil sponges is the calcarean *Raphidonema*, an open cup-shaped sponge that frequently has a knobbly surface. Another, *Barroisia*, has finger-like branches. When broken open, each branch can be seen to have a hollow central cavity surrounded by a ring of chambers. The branches of *Barroisia* represent modules of a kind and some palaeontologists therefore regard this sponge as a colonial animal.

The Cretaceous Chalk of Europe contains a huge number of well-preserved fossil sponges, and the existence of even more in the sea at the time is indicated by numerous flint nodules formed by the precipitation of silica from dissolved demosponge and hexactinellid sponge skeletons. Chalk sponges include cup-shaped forms with a reticulate surface texture, branching forms, and others with stalks and sometimes roots. Among stalked demosponges is the striking *Siphonia tulipa*, the name of the species alluding to its tulip-like appearance.

Corals

People know that corals today construct reefs, which support diverse and colourful communities of other animals. They may also be aware that coral reefs are under threat from human activities, which have increased global levels of carbon dioxide, warming the oceans and causing coral bleaching and death. Dissolved carbon dioxide also increases water acidity, to the detriment of corals forming skeletons. But what exactly are corals? The name coral is applied to species of the phylum Cnidaria that have hard skeletons of calcium carbonate. Other cnidarians include jellyfishes and anemones, which do not have such skeletons. Cnidarians are relatively simple animals lacking the coelom – a fluid-filled body cavity – present in more advanced animals. What they do have is a gut with one opening serving as both mouth and anus.

Calcified skeletons have evolved many times in cnidarians. Thus, corals do not constitute a single taxonomic group: while some corals belong to the class Hydrozoa, the majority are members of the class Anthozoa, and even within anthozoans, corals have evolved several times from anemone-like ancestors during their long evolutionary history. Aside from possessing a hard skeleton, almost all corals have septa. These radial plates correspond to the fleshy mesenteries seen in cnidarians lacking skeletons. It is because they have septa that fossil corals can be distinguished from both sponges and bryozoans. Other components of the skeleton found in many corals include horizontal divisions called tabulae, cyst-like walls known as dissepiments, and a central axial structure. Natural sections of fossil corals visible on rock surfaces or in prepared thin sections are the best way to see these features which, in living corals, are hidden beneath a covering of soft tissues.

Sitting in depressions known as calices at the top of the skeleton are the coral polyps, each equipped with tentacles bearing batteries of stinging cells to capture prey. Consuming zooplankton is not the only way that corals obtain nutrition: many species harbour algal symbionts – zooxanthellae – which furnish the host coral with nutrients including glucose and amino acids produced during photosynthesis in return for shelter. Corals hosting zooxanthellae are known as 'hermatypic'; those without zooxanthellae are called 'ahermatypic'. Because their algal symbionts require light to photosynthesize, hermatypic corals live in shallow waters where they may build reefs. Coral reefs are wave-resistant structures rising significantly above the level of the seafloor and providing complex habitats for a great variety of other organisms. Zooxanthellae have an important role in reef building as they allow their coral symbionts to grow larger, stronger and more rapidly. For instance, branches of the reef-building coral *Acropora* can grow as much as a centimetre each week. Unfortunately, zooxanthellae do not fossilize and their presence in ancient corals must be inferred using other evidence, including the composition of the stable isotopes in the skeleton.

Corals can be solitary or colonial. Depending on species, the single individual in solitary corals varies from cylindrical to an inverted cone or horn- to dome-shaped. The outer bounding walls (epitheca) often lack a soft tissue envelopment when alive. Colonial corals are more varied in form. In many, the modules – corallites – are packed tightly together and consequently are polygonal in plan-view, forming a honeycomb pattern. These 'cerioid' colonies can be lamellar, dome-shaped, plate- or tree-like. In branching corals with so-called fasciculate colonies, the polyps become separated from one another as the tips of the cylindrical branches supporting them diverge. Solitary and colonial corals may both have similar skeletal features such as septa and tabulae, but individual solitary corals are usually somewhat larger in diameter than the equivalent corallites of colonial corals. Hermatypic and ahermatypic species exist in both solitary and colonial corals, although most hermatypic species are colonial.

Corals belonging to three orders are by far the most likely to be found as fossils. These orders are: Rugosa and Tabulata, occurring in rocks of Ordovician–Permian age, and Scleractinia, which range from the Triassic to Recent. Surprisingly, rugose

ABOVE Sectioned fossil of the solitary coral *Amplexizaphrentis* showing the radial septa with one wider gap (cardinal fossula) characteristic of rugose corals. From Matlock Bridge in Derbyshire, England, this Early Carboniferous coral measures 15 mm (½ in) in diameter.

ABOVE RIGHT Fossils of the rugose coral *Dibunophyllum bipartitum* ornament this polished column made from Early Carboniferous Frosterley Marble quarried in County Durham, England.

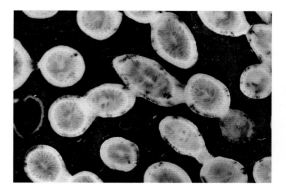 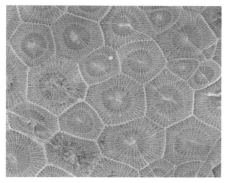

ABOVE Colonial rugose coral *Siphonodendron martini* from the Early Carboniferous of Bristol, England. The rounded cross-section of the corallites is clear in this polished specimen. Field of view is about 5 cm (2 in) wide.

ABOVE RIGHT In the rugose coral *Lithostrotion decipiens* from the Early Carboniferous of Wells in Somerset, England the corallites are tightly packed and have a polygonal cross-sectional shape. Field of view about 6 cm (2¼ in) wide.

and tabulate corals are typically better preserved than the younger scleractinian corals. This is because rugose and tabulate corals have skeletons of calcite, which is more stable through geological time than the overwhelmingly aragonitic skeletons of scleractinian corals. Scleractinian skeletons tend either to be dissolved away, leaving hollow moulds in the rock, or else altered by recrystallization.

Although not always immediately obvious at first sight, the corallites of rugosans are bilaterally symmetrical, making them different from other corals. A line of symmetry passes through a distinctive structure known as the cardinal fossula. Rugose corals can be very abundant in limestones deposited in shallow waters, often in more mud-rich habitats than those inhabited by most corals living today. They occasionally built small mounds but failed to form the large reefs comparable to those of modern scleractinian corals and may well have lacked symbiotic zooxanthellae.

Rugose corals are particularly abundant in the British Carboniferous and include *Amplexizaphrentis* and *Dibunophyllum*. *Amplexizaphrentis* is a classic horn-coral. A few centimetres long, it is believed to have rested on the seafloor, the pointed end buried in mud and the calicular surface – and hence the polyp – elevated above the sediment. Larger in size and more cylindrical in shape, *Dibunophyllum* has an axial structure resembling a cobweb in sectioned specimens. *Dibunophyllum* is abundant in the Frosterley Marble, a black limestone (not a true marble) once quarried in County Durham for use as a monumental stone. Polished examples of Frosterley Marble can be seen in many of the churches of northern England, including Durham Cathedral. Stonemasons in the twelfth to fourteenth centuries used it for small columns, monuments and grave covers, the white corals standing out decoratively against the black rock matrix.

Two closely related rugose genera found in the Carboniferous serve to illustrate the contrast in morphology between colonial corals with different colony-forms. Commonly found in rocks moved by glaciers – glacial erratics – they can be collected on the coast of eastern England in particular. One is *Siphonodendron*, which has a fasciculate colony with cylindrical corallites about 5 mm in diameter. The polyps of *Siphonodendron* colonies would have been isolated from one another during life, each perched on the tip of a branch. The second, *Lithostrotion*, has corallites that are juxtaposed and polygonal in outline, forming a mechanically stronger cerioid colony in which the polyps were potentially better integrated physiologically than those of *Siphonodendron*. The structural and physiological advantages held by *Lithostrotion* over *Siphonodendron* may have been offset by the ability of the latter to cope with polyp smothering by sediment accumulation, a significant problem for corals of all kinds. In *Siphonodendron*, individual polyps could simply push sediment into the gaps between the branches.

Tabulates are a heterogeneous group of exclusively colonial corals characterized by having short or no septa and, in contrast to rugose corals, they often have walls with pores. The Silurian period was the heyday for tabulates – magnificent examples of these corals have been collected from sites in the Welsh Borderlands, the Swedish island of Gotland and the Niagara region of North America, where they often contributed to reef formation. Three common and distinctive tabulate genera are worth highlighting. *Favosites* is known as the honeycomb coral, reflecting its tightly packed polygonal corallites; *Halysites* is referred to as the chain coral on account of its pipe-like corallites linked to form meandering rows; while in *Heliolites* the corallites are set in a matrix of small tubules known as coenenchyme.

The fossil record of the extant scleractinian corals can be traced back some 240 million years to the middle of the Triassic period. Their evolutionary origin has been hotly debated: are they direct descendants of Palaeozoic rugose corals, or did they evolve from a previously soft-bodied group of anthozoans that acquired a calcareous skeleton? A third possibility exists. A few corals of Ordovician age strongly resemble scleractinians. But if these 'scleractiniamorphs' are ancestral to scleractinians, a huge 200-million-year gap in the fossil record must exist.

It did not take long for scleractinians to take up a role of reef building in the Late Triassic. Most evidence points to these corals being hermatypic. Reefs constructed by scleractinians have not always been common: few scleractinian reefs are known in the Late Cretaceous, a time when they were usurped by rudist bivalve reefs, while the Eocene is conspicuously poor in scleractinian reefs before their proliferation in

INVERTEBRATES OF ALL SHAPES AND SIZES • 113

ABOVE Colony of the tabulate coral *Favosites gothlandicus* from the Silurian Wenlock Limestone of Dudley in the West Midlands, England, 9.5 cm (3¾ in) in diameter. The almost circular depression may be where an unpreserved organism fouled the surface of the coral, impeding its growth.

LEFT Colony of *Heliolites megastoma* from the Silurian Wenlock Limestone of Ironbridge in Shropshire, England. This colony is 4 cm (1½ in) across.

BELOW LEFT Detail of the surface of the tabulate coral *Halysites* from the Silurian of Gotland in Sweden. Appropriately dubbed the 'chain coral', the field of view of this image is about 5.5 cm (2 in) across.

the Neogene, culminating in the magnificent but now threatened coral reefs of today. Reefs containing abundant fossils of scleractinians are found in Upper Jurassic rocks across Europe. This includes the so-called Corallian Group of Britain where coral reefs and other beds rich in corals can be found. For example, at Steeple Ashton in Wiltshire more than 40 species of scleractinians have been recorded from a single coral bed. Here and elsewhere in the Jurassic common finds include cerioid colonies of *Isastrea* and *Pseudodiplocoenia* with their star-like corallites, and branching colonies of *Thecosmilia*, along with the closely related but solitary cone-shaped *Montlivaltia*.

Chomatoseris is an unusual Jurassic coral that is shaped like a button, with a flat bottom and convex upper surface. This small solitary coral was apparently auto mobile, having the ability to move slowly over sand on the seafloor, right itself if turned upside down, and escape to the surface if buried. Mobility was achieved in a different way by another scleractinian coral – *Septastrea* – that is found in the Pliocene

RIGHT Fossil of the solitary scleractinian coral *Montlivaltia dispar* from the Jurassic of Nattheim in Germany. The radiating septa visible on the calicular surface of the coral are expressed as parallel lines running down the outside of the skeleton. About 3 cm (1 in) across.

BELOW RIGHT Two specimens of the Pliocene coral *Septastrea marylandica*. Fossils of this scleractinian from Sarasota in Florida, USA evidently lived symbiotically with hermit crabs, which are not fossilized, the coral growing over a gastropod shell, as seen in the 2.5 cm (1 in) tall sectioned specimen on the left.

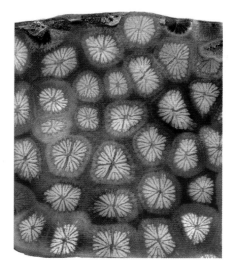

ABOVE Polished section of the Late Jurassic scleractinian coral *Pseudodiplocoenia oblonga*, one of several species referred to as Tisbury starstones in Wiltshire, UK. The polygonal corallites are about 5 mm in diameter.

ABOVE RIGHT Button coral *Chomatoseris bouchardi* from the French Middle Jurassic. About 2 cm (¾ in) in diameter, fossils of this species are also common in some British Middle Jurassic limestones.

of Florida. Colonies of *Septastrea* evidently grew on gastropod shells occupied by hermit crabs. The coral and hermit crab formed a symbiotic association, the active crab preventing accidental burial of the coral, while the coral grew outwards from the aperture of the gastropod shell to construct an enlarged home for the crab. It may also have protected the crab because of its stinging cells and by affording camouflage.

Bryozoans

Try asking 'Alexa' what is a bryozoan and you'll receive an answer laughably wide of the mark. Yet bryozoans constitute a distinct phylum of animals with more than 6,000 species living today, most in the sea but a few in freshwater, and a rich fossil heritage beginning in the Cambrian. Bryozoans form colonies consisting of tubular or box-shaped zooids, each generally less than a millimetre in size but sometimes numbering thousands or even tens of thousands. A good way of recognizing bryozoans is by the presence of an opening in the calcareous skeleton of the zooids through which a retractable crown of tentacles could be protruded to capture plankton. The anatomy of bryozoan tentacle crowns strongly resembles the lophophores of brachiopods (see Chapter 5), and the close relationship between the phyla Bryozoa and Brachiopoda has been corroborated by molecular data, notwithstanding the stark contrast in the appearance of bryozoan colonies and brachiopod shells.

The modular construction of bryozoan colonies has allowed the evolution of a far greater range in shapes than is possible in unitary animals like brachiopods. Modular zooids can be budded in different configurations to produce colonies of widely different form. Common bryozoan colony-forms include branching and patch-like colonies encrusting hard substrates, bushy colonies resembling miniature corals, foliaceous colonies comprising two layers of zooids back-to-back, and lacy, net-like colonies. Encrusting colonies usually remain cemented to their original substrates during fossilization and can be found by carefully searching the surfaces of brachiopods and bivalves with a hand lens. Erect colonies are typically recovered as branch fragments but can still be identified because the diagnostic features present in a single zooid may be sufficient for this purpose. Indeed, zooid morphology is essential to species identification because virtually indistinguishable colony-forms have evolved repeatedly in different groups of bryozoans through geological time.

Bryozoans first become common as fossils in Ordovician rocks. The Cincinnati region of the USA is famed for its rich and well-preserved bryozoans, many belonging to the order Trepostomata, the stony bryozoans. Trepostomes most often have bushy colonies with branches constructed of tightly packed, tubular zooids terminating in polygonal apertures through which the tentacle crowns would have been protruded. Regularly spaced hummocks called monticules may cover branch surfaces. These represent sites where water – filtered of plankton by the zooids – was expelled from the colony. Another order of bryozoans, Fenestrata, becomes commoner later in the Palaeozoic. Most fenestrate bryozoans, which are abundant in the Carboniferous Limestone of Britain, have net-like colonies with narrow branches enclosing holes called fenestrules. Observations on living bryozoans with a similar colony-form show that during feeding the zooids create a flow of water passing in one direction through the fenestrules. In *Archimedes*, the branches are supported by a central screw resembling the water-pump named for the Greek philosopher Archimedes. North American Carboniferous rocks often contain the resistant screw-shaped axes of *Archimedes* that survive after the delicate branches have broken off.

Bryozoans suffered a major extinction at the end of the Permian. No species of fenestrates survived this crisis but trepostomes dwindled on into the Triassic before they too became extinct. Almost all Jurassic bryozoans belong to the related order Cyclostomata, often occurring as encrustations resembling lichens on the surfaces of shells. The Jurassic was marked by the appearance of Cheilostomata, the dominant

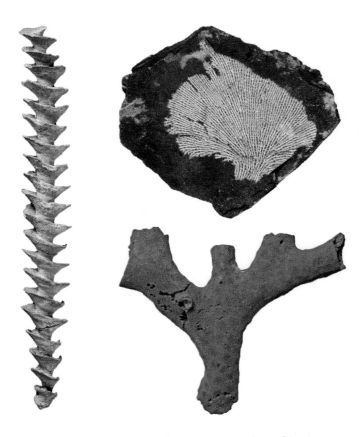

FAR LEFT Colonies of the bryozoan *Archimedes* have a distinctive screw-like axis supporting the branches which are commonly broken off. The axis in this specimen from the Early Carboniferous of Iowa, USA is 18.5 cm (7¼ in) long.

ABOVE LEFT Measuring 6 cm (2¼ in) across, this Early Carboniferous fossil of the bryozoan *Fenestella plebeia* from Flintshire in Wales shows the net-like form of the colony.

LEFT Part of a ramose colony of *Dekayella ulrichi* from the Late Ordovician of Cincinnati, Ohio, USA, one of numerous trepostome bryozoans abounding in these rocks. The fossil measures 6 cm (2¼ in) across.

order of bryozoans in the oceans today. Cheilostomes evolved biomineralized skeletons from a soft-bodied ancestor independently of the more ancient orders mentioned previously. They usually have box-like zooids, in contrast to the tubular zooids of the other bryozoan orders, and their skeletons tend to be more complex, is seen in a particularly attractive group called cribrimorphs.

Polymorphism is almost ubiquitous among bryozoans and is most fully manifested in cheilostomes. Widespread in cheilostomes are polymorphic zooids called avicularia, which received their name because some have a vague resemblance to a beaked bird's head. Avicularia are unable to feed but possess a jaw-like mandible that appears to be defensive, in some species allowing small would-be predators to be captured and held until they die. Mandibles of avicularia evolved by enlargement of the trapdoor-like opercula found in feeding zooids. The majority of cheilostome species possess capsule-like ovicells used to brood embryos before their release as larvae. Fossil cheilostomes show clearly how ovicells evolved from spine-like polymorphs during the mid-Cretaceous at about the same time as avicularia made their debut.

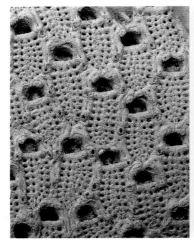

RIGHT Scanning electron microscope images of two cheilostome bryozoans showing details of their intricate skeletons. The image on the near right is of the Late Cretaceous cribrimorph *Pelmatopora* from Rottingdean, East Sussex, England. The far right image shows a Panamanian Plio-Pleistocene colony of *Copidozoum* with small, pointed avicularia and porous, globular ovicells. Fields of view are 2.3 mm for *Pelmatopora*, and 750 microns for *Copidozoum*.

Perhaps the most remarkable of all bryozoans are the lunulite cheilostomes, which have mobile colonies not anchored to a hard substrate. Unlike other bryozoans, the cap-shaped colonies of these bryozoans grow to a regular size, generally a centimetre or so in diameter, their larvae having settled on sand grains or other small grains, often in environments lacking the large substrates required by most other bryozoans. Feeding zooids open only on the convex upper surface of the colony and numerous polymorphs called vibracula (avicularia with hair-like appendages) extend downwards from the colony perimeter to support the colony above sandy or silty sediment. In some species of lunulites these appendages show coordinated movements and act as legs to propel the colony slowly over the seafloor. They also allow colonies that have been accidentally buried to return to the surface. The first lunulites appeared in the Late Cretaceous, becoming commoner through time as various groups of cheilostomes evolved free-living lifestyles.

Fossil bryozoan skeletons can be sufficiently abundant to form limestones, such as the Oligocene Ototara Limestone of Oamaru in

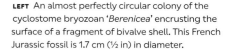

LEFT An almost perfectly circular colony of the cyclostome bryozoan '*Berenicea*' encrusting the surface of a fragment of bivalve shell. This French Jurassic fossil is 1.7 cm (½ in) in diameter.

ABOVE Free-living lunulite bryozoan colonies from the Late Cretaceous Chalk of Rügen, Germany. Each colony is just a few millimetres in diameter.

ABOVE RIGHT Fragments of bryozoan colonies in a Miocene bryozoan limestone from the York Peninsula of South Australia. The field of view is about 3 cm (1 in) across.

New Zealand and several Cenozoic limestones in southern Australia. Oamaru is famed for its white stone buildings, which are made of fine fragments of bryozoan skeletons. The 'corallines' in the Pliocene Coralline Crag of Suffolk, England are in fact bryozoans. Approximately 140 species are present, exhibiting a plethora of different colony-forms. Some belong to genera that today inhabit waters of the Mediterranean or the Atlantic coast of North Africa, attesting to the warmer climate of southern England when they were alive.

Graptolites

A popular prank played by generations of geology students is to use a pencil to draw stick-like marks on pieces of Palaeozoic shale and attempt to pass them off as graptolites. On cursory inspection many graptolites do indeed resemble pencil marks, as reflected in the name, from the Greek *graptos* meaning 'written' and *lithos* meaning 'rock'. However, closer examination reveals jagged edges corresponding to protruding tiny tubes (thecae), each accommodating a single zooid when these colonial animals were alive. Specimens sometimes preserve details such as the stolons that linked the zooids within the colony, but most graptolite fossils are flattened organic films revealing little more than the basic outline of the colony.

Graptolites lacked a biomineralized skeleton and their preservation is due to the resistant organic skeleton, composed mainly of the protein collagen, which was polymerized during fossilization. When viewed at high magnifications in the best-preserved specimens, the surface of the graptolite skeleton is criss-crossed by strips of material which were plastered onto the growing skeleton by the individual zooids. A rod-like structure called the nema forms a kind of 'backbone' running along the branches (stipes) of the colonies.

The soft parts of graptolites are not fossilized and must be inferred using evidence from modern forms. Fortunately – and contrary to what is written in older textbooks – there is one living graptolite genus that can be used for this purpose. *Rhabdopleura* is the only surviving representative of the subclass Graptolithina, which first appeared in the middle Cambrian fossil record and proliferated during Ordovician and Silurian times to total more than 600 genera. *Rhabdopleura* constructs simple, irregular encrusting colonies, but extinct graptolites had much more varied, complex and precisely branched colonies. Nevertheless, it is believed that the zooids of extinct graptolites were like those of *Rhabdopleura*, equipped with two arms bearing tentacles used to feed on plankton.

Graptolites belong to the phylum Hemichordata, which as the name suggests is closely related to the chordates, a phylum that includes animals with backbones (vertebrates). Two main orders of fossil graptolites are recognized: Dendroidea ranging from Cambrian to Carboniferous, and Graptoloidea ranging from Ordovician to Early Devonian. Whereas dendroids formed creeping encrusting or bushy upright colonies attached to firm substrates on the seafloor, the commoner and more diverse graptoloids were planktonic animals. Palaeontologists believe that the first graptoloid, *Rhabdinopora*, evolved from a dendroid, such as *Dictyonema* with numerous branches. A general trend in graptoloid evolution is for the number of branches to be reduced; most Silurian graptoloids consisting of just a single branch. Planktonic graptoloids probably floated in the oceans, moving up and down the water column as they filtered out smaller planktonic food particles. In some genera the colonies had a spiral shape and may have corkscrewed to sweep plankton from a large volume of water as they migrated. Rare aggregations of numerous graptoloids belonging to colonies of a single species united by their earliest formed parts are known as 'synrhabdosomes'. It was once believed that such colonies were attached to a floatation device. Although this theory has been abandoned, how and why synrhabdosomes formed are unanswered questions.

The immense value of graptolites as stratigraphical index fossils in rocks of early Palaeozoic age has long been appreciated. Planktonic graptoloids, by virtue of their lifestyles, were widely distributed in the world's oceans and evolved rapidly, with many species becoming extinct within 2 million years of their first appearance. For example, a scheme of graptolite zones established for the Ordovician of Australasia allows correlation of sedimentary rocks over the entire region. A total of 30 graptoloid zones are recognized, each with an average duration of 1.5 million years.

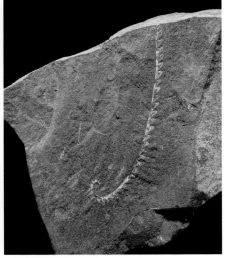

TOP LEFT The Silurian graptolite *Cyrtograptus murchisoni* is characterized by the spiral shape of its branches. The field of view is 3.5 cm (1¼ in) across in this image of a specimen from Czechia.

TOP RIGHT The Silurian graptolite *Monograptus triangulatus*. The saw-tooth-like structures along one side of the branch delineate the skeletons of the individual zooids in this fossil, which is about 3 cm (1 in) long.

LEFT Several colonies of *Dictyonema* from the Early Ordovician of Pedwardine in Herefordshire, England. The rock containing these dendroid graptolites is 7 cm (2¾ in) across.

ABOVE Two colonies of the tuning fork graptolite *Didymograptus* from the Ordovician of Wales. The field of view is about 8 cm (3 in) across.

Two popular sites for collecting graptolites in Britain are Abereiddy Bay near St Davids in Wales, and Dob's Linn near Moffat in Scotland. An old slate quarry at Abereiddy Bay is now flooded by the sea, which is eroding the soft shales of Middle Ordovician age. These shales are almost black and contain profuse graptoloids preserved on bedding planes as pale films. Most belong to the 'tuning fork graptolite' *Didymograptus*, and are variously distorted as the rock was subjected to intense pressure. The oldest rocks at Dob's Linn are slightly younger than those at Abereiddy Bay but are overlain by rocks ranging up into the Silurian. This famous site in southern Scotland was studied by Charles Lapworth (1842–1920), the British geologist who named the Ordovician period. Lapworth recorded changes in the graptoloids present through the sequence of exposed rocks. Dob's Linn is now the 'stratotype' for the boundary between the Ordovician and Silurian periods internationally, the first appearance there of the graptolite *Akidograptus ascensus* marking the base of the Silurian.

Although planktonic graptoloid colonies disappeared from the oceans of the world 400 million years ago, planktonic colonies belonging to unrelated groups inhabit modern oceans, notably salps (phylum Urochordata) and siphonophores (phylum Cnidaria).

'Worms' and conical tubes

The word 'worm' is a loose term applied to long, thin animals belonging to numerous phyla including Platyhelminthes (flatworms), Nemertea (ribbon worms), Annelida (earthworms, ragworms, leeches), Sipunculida (peanut worms) and Phoronida (horseshoe worms). Worms have diverse ecologies: some are parasites, others are carnivores or suspension feeders; some live in the sea, others in freshwater or in the soil. Remarkable fossils of soft-bodied worms have been discovered in Cambrian and younger rocks, but these are far less common as fossils than the biomineralized tubes constructed for protection by a few types of worms.

LEFT Electron microscope image of a small tube, 800 microns in width, constructed by a spirorbid polychaete worm encrusting a shell from Pliocene deposits of Waccamaw, North Carolina, USA.

BELOW LEFT A Jurassic oyster reproducing the topography of an ammonite on which it grew, is itself fouled by an S-shaped serpulid polychaete tubeworm. The fossil is about 6 cm (2¼ in) across.

Fossil tubeworms are abundant in marine rocks of Jurassic to Recent age. Most belong to a family of annelids called Serpulidae. In living serpulids, a crown of appendages consisting of feathery tentacles is protruded from the open end of the calcareous tube for the purpose of suspension feeding but can be retracted into the tube when danger threatens. Serpulid tubes are normally composed of calcite, which explains why they fossilize so well, and can be found attached to shells and rocks, often in dense aggregations. They vary in size and shape – *Spirorbis* has tightly coiled tubes just a few millimetres in diameter, whereas larger serpulid tubes can be straight or sinuous, smooth surfaced or ornamented by one or more ridges running along their lengths. At least 300 species of serpulids are alive today but it is difficult to estimate serpulid diversity in the geological past from their fossilized tubes without the soft parts needed for precise identification.

ABOVE This scanning electron microscope image of *Tentaculites* from the Devonian of Libya shows the characteristic ringed tube measuring 5 mm in length.

Tubeworms of the Palaeozoic were once believed to be serpulids, but new studies have revealed important differences in the structure of their tubes and nearly all of these are now placed in the extinct class Tentaculita. Included in this class is *Tentaculites*, an Ordovician–Devonian genus with long and straight tubes usually ornamented by regularly spaced rings. Whereas *Tentaculites* seems to have been a planktonic animal, the larger tubes of *Cornulites* lived cemented to shells and other hard substrates on the seafloor from the Ordovician to Carboniferous. A similar sessile lifestyle characterized *Microconchus*, which has small spiral tubes routinely misidentified as the annelid *Spirorbis* in the past. *Microconchus* can be found in great abundance in the sediments deposited immediately after some mass extinction events before other suspension feeding animals could gain a foothold.

Finally, conulariids are a fascinating group of some 150 fossil species forming acute four-sided cones. Conulariids grew by adding increments of calcium phosphate skeleton around the wider open end of the cone. First appearing in the Ediacaran and locally abundant in Palaeozoic rocks, the last conulariids are known from the Triassic. Together with some other evidence, the distinctive fourfold symmetry of conulariid tubes points to a surprising affinity with jellyfish. Jellyfish belong to the scyphozoan cnidarians and have an often overlooked 'polypoid' stage in their life cycles when they are sessile and which is often stalk-like as in conulariids.

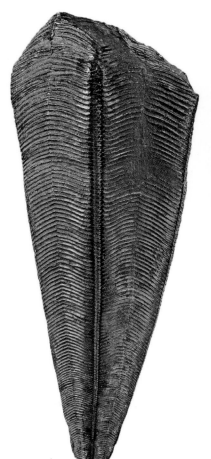

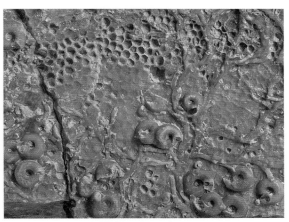

ABOVE Several microconchids with tiny spiral tubes encrusting a coral from the Middle Devonian of Akron in Canada. Resembling spirorbid polychaete tubes, each microconchid is about 3 mm in diameter.

LEFT A beautiful specimen of the conulariid *Paraconularia quadrisulcata*. Measuring 6.7 cm (2½ in) in height, this fossil of a jellyfish relative was collected from Early Carboniferous rocks in Glasgow, Scotland.

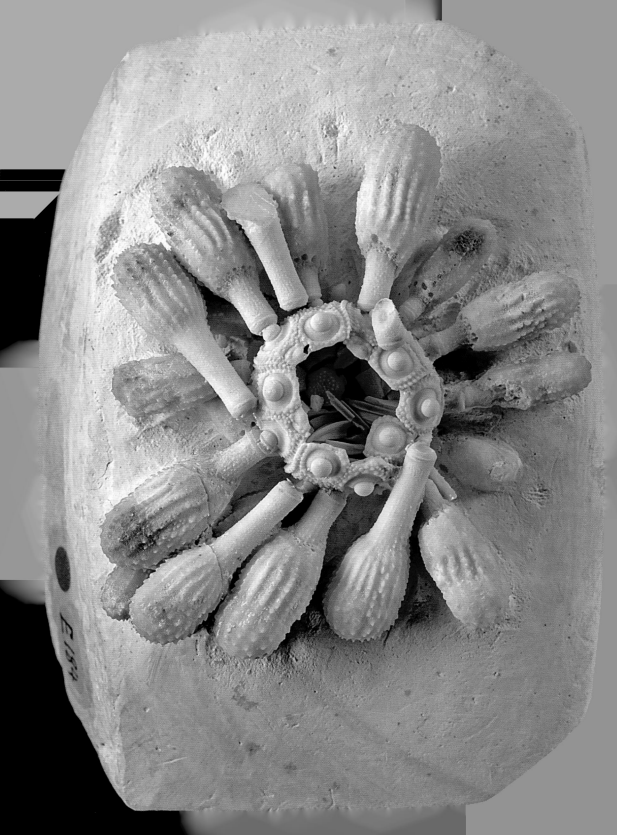

7 The phylum of five: echinoderms

Echinoderms are high in the popularity stakes for fossil collectors, which is not surprising as they include some of the most beautiful and spectacular invertebrate fossils to be found anywhere. Echinoderms are the phylum of five. All extant echinoderms have fivefold radial symmetry – pentamerism – as do most extinct species. There are five living classes of the phylum Echinodermata: Asteroidea (starfishes), Echinoidea (sea urchins), Holothuroidea (sea cucumbers), Ophiuroidea (brittle stars), and Crinoidea (feather stars and sea-lilies). And when the extinct classes are added to the tally, five subphyla of echinoderms are distinguished: Homalozoa, Crinozoa, Asterozoa, Echinozoa and Blastozoa.

The pentaradial symmetry of echinoderms is most clearly seen in the five arms of the common starfish *Asterias rubens*, and in the five sectors containing tube feet in the common sea urchin *Echinus esculentus*. No other animal phylum displays fivefold symmetry. This is not the only unique feature of echinoderms. In addition to features of their soft part anatomy that may not be obvious in fossils, notably a unique system of fluid-filled tubes called the water vascular system used in locomotion, food capture and respiration, echinoderms manufacture skeletons comprising numerous 'ossicles', which are embedded in the body wall. Each ossicle consists of a complex three-dimensional latticework, or 'stereom', made from calcite microcrystals with identical crystallographic alignments. Echinoderms have the capacity to remodel their skeletons as they grow by resorbing parts of the ossicles and adding new calcification elsewhere. Their skeletons are therefore exceptionally dynamic compared with most other invertebrates.

Echinoderm calcite is characterized by a high content of magnesium, making it unstable over geological timescales and prone to alteration. During fossilization, the skeleton commonly 'recrystallizes' and the tiny spaces in echinoderm ossicles become infilled by overgrowths of the skeletal calcite. The entire ossicle transforms into a

OPPOSITE The regular echinoid *Tylocidaris clavigera* from the Late Cretaceous Chalk of England. It is uncommon to find fossils of this echinoid still associated with their articulated club-shaped spines that protected the living animal from predators. Size of block 8 cm (3 in) high.

ABOVE Crinoids and other echinoderms catastrophically buried on the Middle Jurassic seafloor are perfectly preserved on the bedding plane of this rock from Wiltshire, England. Field of view 35 cm (13¾ in) across.

single crystal of calcite, which fractures along cleavage planes, giving it a characteristic spangled appearance. Occasionally, when mud infills the pores, the original lattice structure of the stereom may be preserved in fossils.

It is rare for all the ossicles of an echinoderm to be fossilized together in their original life positions. However, echinoderms can be preserved more-or-less intact when buried quickly before the decomposition of the ligaments, muscles and other soft tissues holding the ossicles together. In some instances, the ossicles interlock and remain together. Following disarticulation, the hundreds or thousands of ossicles in an individual echinoderm are generally scattered by currents. Isolated echinoderm ossicles can prove difficult to identify and fossil collectors tend to focus on Lagerstätten in which articulated skeletons, often containing several species belonging to more than one class, occur in dense aggregations. One such echinoderm bonanza was discovered a few years ago in Middle Jurassic rocks in Wiltshire, England.

All echinoderms live in the sea – none has ever colonized freshwaters. This makes them valuable as indicators of the marine origin of the rocks in which they occur. At the present day, different species of echinoderms can be found in all oceans of the world, from the poles to the equator, and at depths ranging from the intertidal zone to the deep sea. Most fossil echinoderms come from shallow-water sediments deposited on the continental shelf.

Crinoids

In the seventeenth century, when the origin of fossils was still being debated, naturalist Robert Plot (1640–1696) referred to some small fossils from Oxfordshire as 'starstones' on account of their five-pointed star shapes, believing they must in some way be related to the stars in the sky. In fact, starstones are ossicles from the stems of crinoids; when articulated, these ossicles are stacked one on top of the other to form a long stem, tethering the 'business end' of the animal to a substrate.

Crinoids have a long and rich fossil record extending back to at least the Early Ordovician. More than 6,000 fossil species have been described, and there are approximately 600 living species. Crinoid diversity peaked in the Carboniferous, when underwater forests of crinoids carpeted large areas of the shallow seafloor. Similar stem-bearing crinoids – known colloquially as sea-lilies – are uncommon today and nearly all inhabit deeper waters. In shallow waters, including coral reefs, they are replaced by stemless crinoids known as feather stars.

Stemmed crinoids are fixed at the base with a holdfast tethering the stem, which supports the crown. The latter comprises a calyx and multiple arms, usually bearing small branches called pinnules. Holdfasts vary according to species. Some are shaped like low domes; others have branched roots ramifying over the surfaces of hard substrates such as shells and rocks. Stems with crowns often become detached from holdfasts during fossilization, making the identification of holdfasts difficult. A few crinoids, however, have idiosyncratic holdfasts. In the Devonian crinoid *Ancyrocrinus* the holdfast is shaped like a drag anchor, probably securing the crinoid in the sediment in flowing water. Bulbous hollow structures called loboliths in the Devonian crinoid *Scyphocrinites* are believed to have acted as floats. If this theory

LEFT Holdfast of the Middle Devonian crinoid *Ancyrocrinus*. This silicified specimen from Charlestown, Indiana, USA measures 2.5 cm (1 in) in height.

is correct, these crinoids led a planktonic lifestyle, with the stem and crown hanging down from the buoyant lobolith. Other crinoids lacking their own floats, such as *Traumatocrinus* from the Triassic of China and the Jurassic genera *Pentacrinites* and *Seirocrinus*, were also planktonic but these achieved buoyancy by being attached to floating logs. Stunning examples of slabs covered with planktonic *Pentacrinites* and *Seirocrinus* can be seen in European museums. One of the largest, from Holzmaden in Germany, has 141 crinoids attached to a log, some individuals having stems exceeding 10 m (33 ft) in length. Decay of the wood, combined with the ever-increasing burden of attached crinoids and other hitchhiking animals, eventually caused the logs to sink to the muddy seafloor, spelling the end for the planktonic communities they hosted.

RIGHT Individuals of the crinoid *Pentacrinites fossilis* lived attached to floating logs. In this example from the Lower Jurassic of Lyme Regis in Dorset, England, several crinoids can be seen hanging down from a piece of black fossilized wood.

BELOW RIGHT The stems of crinoids after death often disarticulate into individual columnals, such as in these columnals from *Isocrinus psilonoti*, which resemble minute starfish. The field of view is 3 cm (1 in) in this Early Jurassic fossil from Lyme Regis, Dorset, England.

TOP LEFT Columnals of crinoids from the mid Carboniferous Yoredale Shales of Ribblehead in Yorkshire, England, are here strung together in the manner of a rosary. Each columnal is about 1.5 cm (½ in) in diameter.

LEFT Sectioned crinoid stems in an Early Carboniferous limestone from Derbyshire, England. Known as Derbydene Marble, this attractive stone has been used in the floors of buildings such as the Royal Festival Hall on the Southbank, London.

Crinoid stems vary from a few centimetres to several metres in length. An enormous number of disc-shaped ossicles, known as columnals, are present in long stems and these constitute the bulk of the material forming crinoidal limestones, which are especially common in the Carboniferous. The largest columnals are about 5 cm (2 in) in diameter but more often they are nearer 1 cm (½ in) wide. Although most are circular, some are star-shaped as already mentioned, and others are elliptical, the two flat surfaces commonly bearing a radial pattern of ridges that locked into grooves on the opposing columnal. At the centre of the columnal is a hole through which nerves and other tissues passed. This 'axial lumen' can be circular or shaped like a tiny flower with five rounded petals. The hole proved very convenient to early Christian followers of St Cuthbert (634–687) in Northumberland as it allowed crinoid columnals weathered from the local Carboniferous shale and limestone to be threaded together to make rosaries. In fossil folklore these perforated columnals became known as St Cuthbert's Beads.

At the top of the stem is the calyx, or cup, housing most of the organs of the living animal. Crinoid cups are constructed of circlets of plates, the number and arrangement of which vary greatly from one species to another. The simplest cups have a ring of five basal plates followed by an offset ring of five radial plates. More intricate cups containing additional plates tend to be found in Palaeozoic crinoids. The upper surface of the cup, called the tegmen, contains the mouth and anus. The tegmen can be weakly calcified and flexible, or heavily calcified and rigid, as in the Palaeozoic subclass Camerata, in which the cup commonly survives fossilization intact. In some camerates, the tegmen is extended upwards to form a chimney-like anal tube, which would have enabled faeces to be expelled well away from the mouth in the living animal.

Crinoids are suspension feeders using their arms to gather food. The number of arms varies from five to several hundred as a result of multiple branching, and some crinoid arms have small side branches called pinnules. Tube feet on the arms intercept plankton from the seawater. Captured particles are passed into a food groove running down the centre of each arm and conveyed to the mouth by the tube feet and cilia within the food groove.

One fascinating feature of stemmed crinoids is that they encompass a great variety of morphologies and ecologies. Two crinoid genera common in the Jurassic perfectly illustrate this. This first is *Isocrinus*, which is related to the planktonic *Pentacrinites* discussed earlier and shares with it columnals of pentagonal shape, although it was a benthic animal living on the seafloor. The long stem in both genera has occasional columnals bearing root-like cirri. Over 100 arms may be present and, by analogy with living relatives such as *Neocrinus*, these may have been configured to form a parabolic feeding fan, with the tops of the stems bent through 90° in the direction of current flow, the tips of the arms curving upstream, and the mouth facing downstream. Attachment to the seafloor was achieved using cirri in the basal parts of the stem, which could penetrate the sediment or wrap around objects. Remarkably, living crinoids attached in this way have been observed to crawl across the seafloor using their arms to pull the stem along. Unmistakeable fragments of the second

ABOVE A 9.2 cm (3½ in) tall fossil of the Middle Jurassic crinoid *Apiocrinites elegans* from Wiltshire, England, preserves the top of the stem and the cup but not the arms.

LEFT Some crinoids lost their stems and became free-living during evolution. These include the comatulids of which this is an early example from the Middle Jurassic of Wiltshire. The cirri, which would have allowed the living animal to grasp objects on the seafloor, are intact.

genus, *Apiocrinites*, can be seen in the Massangis Limestone from Burgundy, France, which was used to construct the new D-Day memorial at Ver-sur-Mer in Normandy, and more complete specimens of this genus were once collected by the score from the Bradford Clay of Wiltshire, England. The Bradford Clay specimens show *Apiocrinites* to possess a short, straight stem, the disc-shaped columnals broadening in width at the top of the stem where there is a smooth transition to a large, barrel-like cup. *Apiocrinites* lived permanently cemented to a hard surface via a thick and strong holdfast. Unlike *Isocrinus*, there are no cirri. The 10 arms are short, delicate and widely separated, unsuited to forming the kind of parabolic feeding fan found in *Isocrinus*.

Crinoids without stems have evolved on multiple occasions from stemmed ancestors. The most diverse belong to the order Comatulida, the feather stars, which first appeared in the Early Jurassic and are represented by more than 500 species in the sea today. In place of the stem, comatulids have a single plate called the centrodorsal at the base of the cup. Cirri, which are used by comatulids to grasp objects, are borne on the centrodorsal. The most easily identifiable fossils of comatulids are the bowl-shaped centrodorsals with tell-tale sockets where the cirri were once attached. Some modern species of comatulids can swim by moving alternate arms up and down in a coordinated and graceful motion. A popular hypothesis is that the switch in dominance from stemmed crinoids to comatulids through geological time reflects increasing pressure from predators, with comatulids better able to escape predation because of their ability to swim, and also to hide in crevices during the daytime before emerging to feed at night.

ABOVE Aggregation of the Late Cretaceous stemless crinoid *Uintacrinus socialis* from Elkader in Kansas, USA. Some of the boundaries between the plates of the cup have been emphasised using ink. The rock is 29 cm (11½ in) across.

The largest stemless crinoid – not a comatulid – is a Late Cretaceous species called *Uintacrinus socialis*, the genus name derived from the Uinta Mountains in western USA, and the species name from the occurrence of mass accumulations of this crinoid. Large slabs from the Niobrara Chalk of Kansas are totally covered by *Uintacrinus* cups with attached arms. The bulbous cups are up to 7.5 cm (3 in) in diameter and consist of numerous polygonal plates, whereas the 10 arms can reach 125 cm (49¼ in) in length. *Uintacrinus* was once believed to be planktonic but has been reinterpreted as living in dense aggregations on the seafloor, the cup partly buried in sediment and the arms held vertically into the water above to capture food using the tube feet.

Echinoids

Echinoids – popularly known as sea urchins – are another collectors' favourite; their alternative name is sea hedgehogs, alluding to the covering of hedgehog-like spines. The spiny character of echinoids is not obvious from the bare examples bought from seaside shops to enliven bathroom displays, as echinoid spines routinely drop-off after death owing to decomposition of the organic tissues that join them to the main

LEFT The irregular echinoid *Clypeus plotii* gives it name to the Clypeus Grit Member, a stratigraphical unit in the Middle Jurassic of the Cotswolds. Specimens collected by local people are known as 'poundstones' or 'Chedworth buns'. This example from Notgrove in Gloucestershire, England measures 11 cm (4½ in) across.

part of the animal. Likewise, fossil echinoids are typically discovered without their spines, which are sometimes found loose in the surrounding sediments.

More than 1,000 species of living echinoids have been recognized and they may be more diverse today than at any time in the geological past. As a group, echinoids are distributed from the shallows of the intertidal zone to the depths of the abyss, and at all latitudes between the tropics and the poles. Although the oldest known indisputable echinoids date from the Early Silurian, they are uncommon fossils throughout the remaining nearly 200 million years of the Palaeozoic era. Subsequent evolutionary radiations and changes in their morphology and ecology, however, resulted in echinoids becoming one of the most prominent fossil groups in Mesozoic and Cenozoic rocks. Echinoid skeletons, known as 'tests', with their distinctive star-shaped markings, are exquisite fossils that have attracted human interest for thousands of years. Indeed, Neolithic graves have been found containing specimens of the Cretaceous echinoid *Echinocorys* arranged around the skeletons, probably not just as decorations but indicating their symbolic or spiritual significance.

Echinoid tests are typically constructed of double columns of plates arranged in 10 radial sectors, five so-called ambulacra alternating with five interambulacra. The ambulacral plates contain pores where the tube feet, which are employed for feeding, moving and respiration, are located. Embellishing the broader interambulacral plates are tubercles, which are the articulating bases of spines attached by muscles and used in locomotion as well as defence. Ambulacra and interambulacra converge on an apical disc at the top of the test, and on the peristome on the underside where the mouth is situated. The apical disc consists of a series of plates, one of which – the madreporite – has minute pores connecting to the water vascular system within the test. The periproct marking the position of the anus may also be located in the apical disc, although this is not always the case (p. 137).

The taxonomy of echinoids is complex and will not be covered here, but their division into two main groups – 'regulars' and 'irregulars' – is important as these groups are both ecologically and morphologically distinct. The tests of regular echinoids show only pentaradiate symmetry, but in irregular echinoids a secondary bilateral symmetry is superimposed on the pentaradiate pattern. Whereas the tests of regular echinoids tend to be globular and circular in plan-view, those of irregular echinoids are often flattened and, in plan-view, may be elongated along the plane of bilateral symmetry. Regular echinoids have five strong teeth, which they use to feed on algae, sessile animals and carrion. Some scrape hard surfaces such as rocks and shells, leaving distinctive gouge marks (see Chapter 13). In contrast, irregular echinoids bulk process sediments to extract organic particles, a method of deposit-feeding that often entails burrowing. They are more diverse and varied than regular echinoids.

The earliest echinoids are regulars, with irregulars first appearing in the Early Jurassic. Regular echinoids were undoubtedly rarer in Palaeozoic seas than they are today, but their scarcity as fossils is also due in part to the loosely joined plates of their flexible tests being easily disarticulated. Rapid burial by mud was needed to preserve intact tests, albeit crushed, along with articulated spines, as in the Carboniferous genus *Archaeocidaris*. The more rigid tests of Mesozoic and Cenozoic regular echinoids

ABOVE The spines of the echinoid *Archaeocidaris brownwoodensis* can be seen radiating out from the crushed test in this exceptional specimen from the Late Carboniferous of Texas, USA on a slab of rock 18.5 cm (7¼ in) across.

ABOVE This view of the upper surface of the Late Jurassic irregular echinoid *Pygaster umbrella* from near Oxford, England shows the keyhole-shaped periproct which contained the anus. The test is 7.8 cm (3 in) across.

ABOVE RIGHT In the irregular echinoid *Holectypus depressus* the periproct is on the underside of the test and hence not visible in this view of the upper surface. This fossil from the Middle Jurassic of Northamptonshire, England is 4.8 cm (1¾ in) across.

survived better. Their spines were large and varied in shape: some flattened, others club-like or extravagantly flared at their ends. Impressive finds of *Tylocidaris* (p. 16) from the Cretaceous Chalk show just how large these defensive spines were relative to the tests of the animals. Isolated spines from this and similar echinoids are robust and can be common fossils in Jurassic–Recent rocks. Disarticulated interambulacral plates are also frequently found, pentagonal in shape and bearing a spine base curiously resembling a human nipple with a central boss surrounded by a dimpled areola.

Echinoids of Jurassic age neatly illustrate evolutionary stages from a pentaradiate regular ancestor to irregular echinoids exhibiting strong bilateral symmetry. One of the key modifications was the 'migration' of the periproct (and hence anus) away from a central position on the apical disc at the top of the test, and onto the side or lower surface of the test. In *Pygaster* the periproct is elongated away from the apical disc, although still in contact with it, on the upper surface of the test. However, in *Holectypus* it is separated from the apical disc and positioned on the underside of the test, placing both the anus and mouth on the lower surface of the test, in marked contrast to all regular echinoids where they are on opposite surfaces.

Heart urchins, sea biscuits and sand dollars are the dominant living groups of irregular sea urchins. As the name suggests, heart urchins have tests indented at one

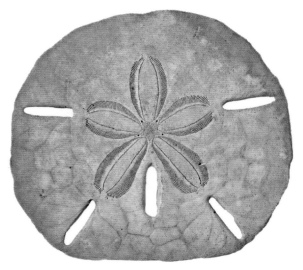

ABOVE The 'heart urchin' *Micraster coranguinum* is very common in the Late Cretaceous Chalk of northern Europe. This example is about 6 cm (2¼ in) in height.

ABOVE RIGHT Five slit-like lunules and a 'petal' consisting of the five ambulacra are conspicuous in this Pleistocene test of the sand dollar *Mellita quinquiesperforata* from Charleston, South Carolina, USA, which measures 7 cm (2¾ in) across.

end to resemble a heart in plan-view. Delicate tests of the living 'common heart urchin' *Echinocardium cordatum* can be found washed up on beaches around the northeastern Atlantic. This animal burrows to depths of up to 15 cm (6 in) in sand, using the fine spines covering the test to create a water-filled space necessary for the buried animal to respire. The best-known fossil heart urchin is the Late Cretaceous–Paleocene genus *Micraster*, which is abundant in the Chalk of Europe, and has been the subject of numerous evolutionary studies. The ambulacra of *Micraster* form a distinctive petaloid structure on the upper surface of the test. Although specimens of *Micraster* in the Chalk are occasionally crushed or fractured, nearly all tests are complete. Yet these specimens are commonly encrusted by suspension-feeding animals that can only have colonized tests exposed on the seafloor. After death the echinoids were apparently exhumed from their burrows, perhaps by erosion of the soft chalk sediment, and were able to survive for long enough to be encrusted.

Sand dollars become increasingly common as fossils from the Eocene onwards, and are often found in dense accumulations, complete or as fragments. As in a dollar coin, they are disc-like and thin, the upper surface of the test marked by petaloid ambulacra like that of *Micraster*. Support structures buttress the interior of the test, adding to the resilience and fossilization potential of the plates. Sand dollars live immediately beneath the surface in fine to medium sands, feeding on fine particles in the sand passing over the surface of the test. Uniquely among echinoids, many species have slit-like perforations in the test called lunules that act as pressure release valves to stabilize the animals during storm surges. In morphology and ecology, sand dollars represent the furthest departure from primitive regular echinoids in morphology and ecology, none more so than the extant genus *Dendraster*, which angles its test obliquely out of the sediment to feed on plankton in the water above.

Asteroids, ophiuroids and holothurians

The three remaining classes of extant echinoderms can all be found as fossils but far less frequently than crinoids and echinoids. Asteroids and ophiuroids typically have small, separated ossicles that become dissociated and disperse when the animals die and the soft tissues containing them decompose. Preservation of articulated fossils of these echinoderms demands very rapid burial and no subsequent disturbance by animals burrowing through the sediment. The ossicles of holothurians (sea cucumbers) are generally reduced to minute spicules, often in the shape of perforated wheels or anchors less than a millimetre in diameter, which are very easy to overlook.

Asteroids – starfishes in the strict sense – are represented by some 1,900 living species. They can have as few as 5 arms or as many as 50 arms (as in the Antarctic species *Labidiaster annulatus*), while the largest species of asteroid is close to a metre in diameter. Most asteroids are predators, but some are suspension feeders or consume organic detritus. All these feeding ecologies may have been present in Palaeozoic species too, although later species not only had the ability to swallow larger items of prey but could also evert their stomachs during digestion. The oldest fossil believed to be a starfish was found in rocks of Early Ordovician age. By the Devonian, asteroids had diversified, as is evident in the famous Hunsrück Slate of Bundenbach, Germany in which asteroids and a variety of other animals were rapidly buried in mud. Early mineralization in this low oxygen setting quickly encased the fossils in pyrite, resulting in exceptional preservation of entire asteroids, including both ossicles and soft parts. Chalk formed 300 million years later, in the Late Cretaceous, has long been a source for articulated asteroids such as *Nymphaster*, although scattered ossicles are far more commonly found. Most Chalk asteroids belong to the order Valvatida and have large marginal ossicles outlining the body, which are distinctive fossils.

Ophiuroids (brittle stars) are often mistaken for the closely related starfishes but differ in having arms sharply demarcated from the central body disc and characteristically long, thin, whip-like, flexible, and able to be bent into sinuous

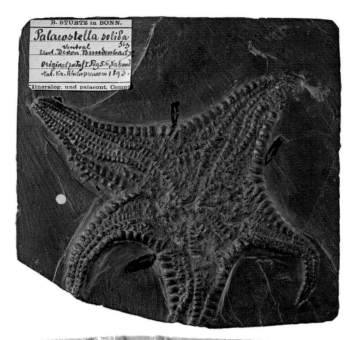

TOP RIGHT Pyritized asteroid *Palaestella solida* from the Early Devonian of Bundenbach in Germany on a slab of Hunsrück Slate measuring 13.5 cm (5¼ in) across.

RIGHT This exquisite fossil of the asteroid *Nymphaster coombii* from the Late Cretaceous of Sussex, England sits on a chalk block 9.5 cm (3¾ in) across.

shapes. More than 2,000 living species of ophiuroids have been described, making them slightly more diverse than asteroids. Their fossil record extends back to the Early Ordovician but is meagre. One bed famed for its fossil ophiuroids is the misnamed Starfish Bed in lower Jurassic rocks of Eype in Dorset, England. Belonging to a species now called *Palaeocoma milleri*, the Eype ophiuroids are believed to have been buried in a single event, possibly a storm or a tsunami. The arms are about 5 cm (2 in) long and tend to be bent in varying patterns, giving the appearance of animals writhing in their death throes.

Holothurians are close relatives of echinoids from which they diverged, probably in the Ordovician. Unfortunately, complete fossils of holothurians, in contrast to isolated spicules, are known from a mere handful of localities around the world. One

ABOVE Several overlapping individuals of the ophiuroid *Palaeocoma milleri* are visible on this 32 cm (12½ in) wide block of Early Jurassic rock from Thorncombe Beacon in Dorset, England.

of these is Collbató in northern Spain, where holothurians occur in a fine-grained, laminated limestone deposited in a tranquil environment during the Triassic. At less than 6 cm (2½ in) long, these fossils have spindle- or sausage-shaped bodies, which are covered by imbricated plates and have a ring of plates surrounding the mouth at one end.

Extinct classes

There are no fewer than 15 extinct classes of echinoderms, all of which are confined to the Palaeozoic. Their existence reflects a seemingly greater variety of morphologies present, some not pentaradial, during the early evolutionary history of echinoderms. It is almost as though echinoderms 'experimented' with numerous different body plans of which only a few, those of the five extant classes, were ultimately successful.

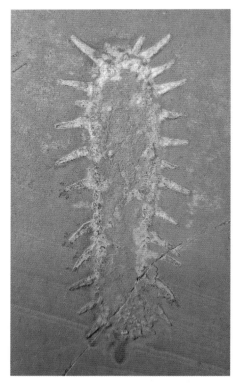

ABOVE A rare example of a fossil sea cucumber measuring 14 cm (5½ in) in length, *Oneirophantites tarragonensis* comes from Middle Triassic rocks in Tarragona, Catalonia, Spain.

Edrioasteroids are an intriguing class of Cambrian–Permian echinoderms vaguely reminiscent of Celtic brooches in appearance. They lived attached to hard surfaces, such as rocks and brachiopod shells, feeding on particles of plankton, and had dome-shaped thecae generally about a centimetre in diameter. As with other echinoderms, rapid burial was necessary for their preservation but even so, the thecae are found collapsed and flattened with the numerous plates somewhat displaced. Five ambulacra radiate from the peristome and mouth at the centre of the theca. Four of the ambulacra are typically curved in the same direction and one in the opposite direction, giving one widened interambulacrum containing the anus. Spectacular aggregations of edrioasteroids are occasionally found, as in the Carboniferous of Cumbria. Edrioasteroids lived here in an environment of unidirectional current flow and apparently swivelled as they grew so that the anus was in the most sanitary arrangement downstream of the mouth.

Cystoids and blastoids are closely related classes of the subphylum Blastozoa. Both first appeared in the Ordovician, with cystoids becoming extinct in the Devonian and

ABOVE Attached to this fragment of brachiopod shell is the edrioasteroid *Carynella pilea*, 1.2 cm (½ in) in diameter. The fossil is from the Late Ordovician of Kentucky, USA.

ABOVE RIGHT Numerous individuals of different sizes belonging to the edrioasteroid species *Stalticodiscus milleri* are attached to the surface of this broken concretion measuring 15 cm (6 in) in height from the Early Carboniferous of Penruddock in Cumbria, England.

blastoids in the Permian. The main body of these echinoderms, the theca, is roughly ovoidal in shape. In both groups it is formed of interlocking plates, producing a rigid structure with a high preservation potential. Attached to the base of the theca in many species is a short stem consisting of stacked ossicles and resembling that of a crinoid. Unlike crinoids, however, cystoids and blastoids lack arms, instead capturing food using long and delicate appendages called brachioles, which are not often fossilized. The theca of cystoids is pierced by a variety of pores, some thought to be respiratory in function, while blastoids have complex internal systems of tubes called hydrospires, also believed to function in respiration. In blastoids, five ambulacra pass down the sides of the theca from the summit where five pores surround a central mouth.

Among the commonest genera of these two echinoderm classes are the cystoid *Echinosphaerites* and the blastoid *Deltoblastus*. Thecae of *Echinosphaerites* are almost spherical, smooth and sometimes sufficiently abundant to be limestone-forming. When broken open they are often seen to be partly or completely filled with inwardly growing radial crystals of calcite. For this reason, Swedish examples of *Echinosphaerites*

RIGHT This fossil of the cystoid *Apiocystites pentremitoides* is 2.6 cm (1 in) long and shows the theca and short, segmented stem. The specimen comes from the Silurian Wenlock Limestone of Dudley.

BELOW Eleven fossils of the blastoid *Deltablastus jonkeri* from the Late Permian of Timor, Indonesia, each measuring 1–2 cm (¼–¾ in) in height.

LEFT Preserved as a mould after dissolution of the calcite skeleton, this is the homalozoan *Cothurnocystis elizae* from the Late Ordovician of Ayrshire in Scotland. The tail-like appendage of this 2.7 cm (1 in) wide individual emerges from the boot-shaped body.

are known in folklore as 'crystal apples'. *Deltoblastus* occurs profusely in the Permian rocks of Timor, Indonesia, and is one of the last surviving blastoid genera. The five ambulacra with their slit-like pores are separated by ridges called deltoid plates on the ovoidal theca.

Finally, there are the Homalozoa, once referred to as carpoids, a subphylum exhibiting primitive features helpful in linking echinoderms to other phyla. These bizarre Cambrian–Devonian fossils have attracted great controversy over the years, with some palaeontologists contending that the appendage on the main plated body (theca) was a stem and others that it was an arm, while pores between the plates of the theca have been variously interpreted as openings for the mouth or for the anus. Despite lacking the pentaradial symmetry seen in other echinoderms, the plates of homalozoans are composed of calcite with the porous stereom structure typical for the phylum. Whereas many homalozoans resemble tiny tennis rackets, *Corthurnocystis* looks more like a Medieval boot with a pointed toe. Slit-shaped openings in the theca of *Corthurnocystis* have been interpreted as outlets, or alternatively, inlets for feeding and respiratory currents and may be the equivalent of the gill slits found in vertebrates. While there are disagreements about numerous aspects of homalozoans, it is widely accepted that when alive these peculiar asymmetrical animals rested flat on one side, with the appendage possibly being used to drag the body over the seafloor and as a feeding organ.

8 Fishes and amphibians: from fins to feet

The following three chapters concern animals with backbones classified in the subphylum Vertebrata, otherwise known as Craniata. Vertebrates consist of fishes, amphibians, reptiles, birds and mammals. By definition, all vertebrates possess a vertebral column composed of vertebrae. These vertebrae are usually made of bone, and form part of an assortment of other bones typically present, such as ribs, limb and skull bones. Their teeth also fossilize because, like bones, teeth are composed predominantly of the calcium phosphate mineral hydroxyapatite, which survives well through geological time. Consequently, although less numerous than invertebrates, vertebrates are common fossils, particularly in Devonian and younger rocks.

Over 66,000 species of vertebrates are alive today, roughly half of which are fishes, that is aquatic vertebrates possessing gills but lacking limbs with digits. From an evolutionary standpoint, however, fishes do not represent a natural biological group. Instead, they are what remains of the vertebrates after tetrapods (amphibians, reptiles, birds and mammals) have been subtracted, and are therefore paraphyletic. Taking vertebrates in totality, modern phylogenetic classification divides them first into two major groups: those lacking jaws (the paraphyletic Agnatha) and those with jaws (Gnathostomata). Gnathostomes consist of two major classes of extant fishes – Chondrichthyes (cartilaginous fishes, including sharks) and Osteichthyes (ray-finned and lobe-finned bony fishes) – plus the Tetrapoda, which themselves are split into three classes – Amphibia, Sauropsida (reptiles and birds) and Synapsida (mammals). Two additional classes of extinct gnathostome fishes important in the Palaeozoic are acanthodians and placoderms.

This chapter begins with agnathans and ends with amphibians. It describes a selection of the commoner and more important fossils, including those revealing stages in the critical transition from fully aquatic fishes with gills and fins, to land-dwelling amphibians with lungs and feet.

OPPOSITE Late Miocene frog *Pelophylax pueyoi* on a piece of rock 16 cm (6¼ in) across from Teruel in northeast Spain. The dark outline of the body is preserved in addition to the fragile bones.

Agnathan fishes

Only two groups of jawless agnathan fishes are alive today: lampreys and hagfish. Both have eel-shaped bodies but lack bones and are not well-represented in the fossil record, although the teeth of a related group – conodonts – are abundant microfossils (see Chapter 12). Several Cambrian soft-bodied fossils with flattened fish-like bodies have been interpreted as early vertebrates but their affinities have not been accepted universally. By the end of the Cambrian, fossils are known with hard parts made of dentine that can be more securely identified as vertebrates, such as *Anatolepis*, the oldest known agnathan.

The evolution of bone led to an increase in the diversity of agnathans as well as their representation in the fossil record. Bone is a remarkable material. In addition to the mineral component, bone contains the protein collagen, and is provisioned with spaces to accommodate living cells and canals for blood vessels. Vertebrates possess two types of bones: endoskeletal bones formed deep within the body, and dermal bones formed immediately below the skin. Agnathans lack a bony endoskeleton but many of the extinct groups found as fossils have a well-developed dermal skeleton consisting of bony scales, plates or denticles.

The Devonian period has long been labelled 'The Age of Fishes'. Many different types of fishes first become abundant in Devonian rocks, even though fossil finds show that their evolutionary roots extend further back in geological time. The Devonian in the British Isles is mainly represented by the Old Red Sandstone, a sequence of non-marine sandstones, conglomerates and mudstones. This includes fish-bearing deposits formed in freshwater environments. One of these fishes is the agnathan *Cephalaspis*. Measuring up to 30 cm (12 in) in length, *Cephalaspis* is heavily armoured with a large head shield and a body covered by dermal scales. A pair of closely spaced holes that accommodated the eyes are present on the top of the head shield, while the mouth

RIGHT This specimen of the agnathan fish *Cephalaspis lyelli* comes from the Devonian 'Lower Old Red Sandstone' of Glammis in Forfar, Scotland. Measuring 21 cm (8 in) long, it is the type specimen of the species originally described by the renowned naturalist Louis Agassiz (1807–1873).

and gill openings are situated on the underside. *Cephalaspis* was a bottom feeder, probably using movements of its headshield to stir up sediment containing small-sized prey. Without jaws, feeding on larger prey would have been difficult. Another agnathan called *Pteraspis* lived in the sea, in contrast to the freshwater *Cephalaspis*, and is believed to have been a more active swimmer, perhaps feeding on plankton near the surface of the ocean. At the front of the head shield in *Pteraspis* is a pointed rostrum, which may have had a defensive function.

Placoderm and acanthodian fishes

These two extinct groups of jawed fishes are well represented in the Devonian, although both have longer geological ranges, with placoderms also being recorded in the Silurian while acanthodians range from the Ordovician to the Early Permian. Acanthodians, colloquially known as spiny sharks, are among the earliest vertebrates with jaws. Although their bodies have the general shape of a shark, with a heterocercal tail in which the upper lobe is larger than the lower lobe, they are not true sharks. The descriptor 'spiny' comes from the presence on the front-facing edge of the fin of a reinforcing dentine spine. Acanthodian spines, as well as the tiny but distinctive diamond-shaped scales that cover the body, are usually all that remains in the fossil record, except for occasional specimens preserved as shadow-like impressions in the rock. Many acanthodians were small and lived in freshwater environments. Some were predators, others were filter feeders consuming small plankton.

The heavily armoured placoderms colonized both marine and freshwater environments. With approximately 350 genera recognized to date, they succeeded in becoming very diverse during their geologically brief period of existence. The bodies of placoderms are covered with bony plates, which are tough and readily fossilized. Two genera serve to illustrate placoderm variety: *Bothriolepis* and *Dunkleosteus*. The peculiar

LEFT Preserved in a concretion from the Devonian of Banffshire, Scotland, this small fossil is the acanthodian fish *Cheiracanthus murchisoni*, measuring 15 cm (6 in) long.

RIGHT *Dunkleosteus terrelli* was a big fish. The dorsal trunk plate (top) and head (bottom), which is almost 50 cm (19½ in) wide, are visible in this example from the Late Devonian of Cleveland, Ohio, USA.

BELOW Fossil of the peculiar placoderm fish *Bothriolepis canadensis* of Late Devonian age. This example is about 6 cm (2½ in) in length.

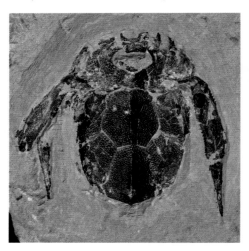

genus *Bothriolepis* was mistaken for a tortoise when first found in 1843 by Abraham Gesner in Devonian rocks at Miguasha, Quebec, Canada. This error is understandable given the box-like bony head and thorax that together resemble a chelonian carapace. *Bothriolepis* also has a pair of long pectoral fins, unusually for placoderms, encased in bone, and a tail, although the latter is less often preserved. Most of the 60 known species of *Bothriolepis* lived in freshwater. *Dunkleosteus* is a different kettle of placoderm. Famous for its huge size, early estimates put the largest species of this Devonian placoderm at 8-9 m (26-29½ ft) in length, although new studies suggest a more compact body less than 4 m (13 ft) long. However, as only the front end of the animal is fossilized, the full size of the animal must be estimated. The massive jaws equipped with sharp bony plates acting like teeth made *Dunkleosteus* a formidable predator, and evidence from puncture wounds piercing the 5-cm (2-in) -thick dermal armour of some fossils of *Dunkleosteus* suggest that it was not uncommon for individuals to attack others of the same species.

Cartilaginous fishes

While the sighting by swimmers of a living shark triggers terror, the discovery of a fossil shark's tooth on the adjoining beach elicits delight, especially if the fossil is an enormous 'megalodon' tooth. Sharks belong to the class Chondrichthyes, which also includes rays and some other fish with skeletons made predominantly of cartilage rather than bone. Although the cartilage of sharks usually has a thin outer layer composed of tiny crystals of hydroxyapatite, this seldom survives to be fossilized and most shark fossils consist of isolated teeth and, less often, fin spines. The pristine, shiny appearance of shark teeth testifies to their resilience. Given the fact that sharks are conveyor-belt tooth factories, growing and shedding thousands during their lifetimes, it is surprising that fossil shark teeth are not far commoner. Part of the reason is that bacteria may infiltrate between the minute crystals of dead shark teeth causing degradation, but as a lot of sharks possess small teeth these can easily go undetected. Sieving of sediment is a good way of finding small shark teeth.

Sharks and rays living today are classified as neoselachians, fossils of which did not become common until the Jurassic – nearly all shark teeth gathered by the devoted legions of specialist collectors are Mesozoic or Cenozoic in age. Neoselachians were preceded by several mostly Palaeozoic groups of sharks and, indeed, sharks are among the most ancient groups of fishes still alive today. Their ancestry may lie within the acanthodians described in the last section. Hybodonts are one of the better-known groups of extinct sharks, ranging from the latest Devonian to the Cretaceous and once living in the sea as well as in rivers. They are distinguished by having two dorsal fins, each of which has a large, gently curved spine.

Two bizarre Palaeozoic sharks – *Stethacanthus* and *Helicoprion* – deserve to be mentioned. The Late Devonian–Carboniferous genus *Stethacanthus* has an unusual anvil-shaped 'brush' on its back where the dorsal fin would normally be situated. This

RIGHT Measuring 45 cm (17¾ in) in length, this is a long spine from the dorsal fin of the shark *Hybodus delabechei*, which was collected from the Early Jurassic rocks of Lyme Regis, Dorset, England.

RIGHT Cast of the spiral tooth whorl of *Helicoprion nevadensis*, a Permian shark from Pershing County in Nevada, USA. Viewed obliquely, this specimen is 14 cm (5½ in) across.

consists of numerous tubes of calcified cartilage and the top is covered with forward-facing denticles. If the claim that the brush is only present in mature males is correct, it may have had a role in courtship or mating. Equally striking is the peculiar tooth whorl of the Permian genus *Helicoprion*. Superficially resembling an ammonite shell, the tooth whorl was carried on the lower jaw and could consist of more than 150 individual teeth arranged in a logarithmic spiral. It seems to represent an adaptation for a particular mode of feeding on soft-bodied prey.

Teeth of the geologically younger neosalachian sharks exhibit a range of shapes from triangular to spear-shaped. Some have a pointed main cusp with smaller cusps ('cusplets') on either side, while others possess a row of about 10 cusps of decreasing size. The cutting edges of fossil shark teeth can be very sharp, capable of drawing blood if handled carelessly, and are often finely serrated. If proof were needed, these traits amply demonstrate their predatory lifestyles, with the serrations reminiscent of the similar structures on steak knives, which make them so effective in cutting through meat. Other sharks, however, have flat teeth adapted to crushing shelled prey such as molluscs. Many rays, also neoselachians, are commonly found as fossils. They have distinctive plate-like teeth used for crushing hard-shelled prey. In 1695, Hans Sloane (1660–1753) whose collections were to form the basis of the British Museum, described a fragmentary fossil from Maryland, USA. Sloane showed remarkable insight in identifying this fossil as the toothplate of a stingray, living examples of which he had observed in the seas around Jamaica a few years earlier when he was a physician aboard HMS *Assistance*.

At roughly three times the length of the great white shark, *Otodus megalodon* was a shark of immense size. Based on its huge teeth, which can reach 18 cm (7 in) in length, various size estimates have been proposed for the length of the living animal.

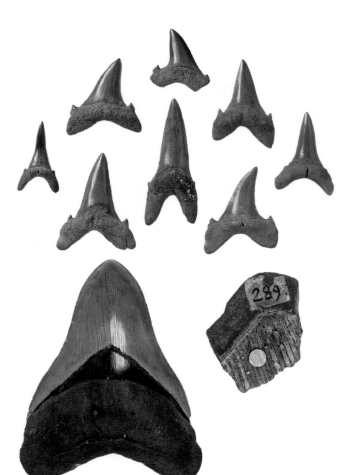

TOP LEFT Eight teeth of *Odontaspis robusta*, a sand tiger shark. These examples are from the Early Eocene of Abbey Wood, Kent, England. The field of view is 9 cm (3½ in) across.

FAR LEFT The teeth of the famous megalodon shark, *Otodus megalodon*, are often over 8 cm (3 in) in length. This example was collected from Miocene deposits in Virginia, USA.

LEFT Fragment, measuring just under 2 cm (¾ in) across, of the tooth of a fossil stingray from Maryland, USA. This specimen was described by Sir Hans Sloane in the *Philosophical Transactions of the Royal Society* for 1695–7.

A conservative guess is a staggering 15 m (49 ft). Specimens of this Miocene–Pliocene leviathan have been found in rocks from many parts of the world, where it reigned as an apex predator probably subsisting on a diet of marine mammals, including small whales.

Ray-finned bony fishes

The great majority of fishes living today and most of those eaten by humans or kept as pets in fish tanks are actinopterygians, ray-finned bony fish with flexible fin rays. These fishes dominate in marine and freshwater habitats and can be traced in the fossil record all the way back to the Devonian. The most successful actinopterygians are a subgroup called teleosts in which the jaws are able to be protruded outwards to allow prey to be grabbed and pulled towards the mouth.

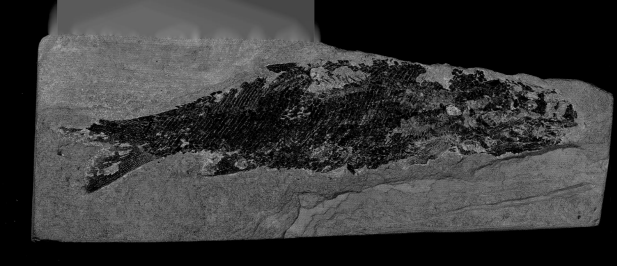

ABOVE Ray-finned fish *Palaeoniscum freieslebeni* from Permian rocks near Middleridge, County Durham, England. This example measures about 20 cm (8 in) in length.

One of the most familiar Palaeozoic actinopterygians is *Palaeoniscum*. Flattened but complete specimens of *Palaeoniscum freieslebeni*, are common in the Permian Kupferschiefer of Germany and its equivalent, the Marl Slate of northern England. Deposited near to the coastline in shallow water, these sediments are carbon-rich (bituminous) shales that accumulated on a low-oxygen seafloor, which favoured the preservation of *Palaeoniscum* and other organic remains. Unlike many later actinopterygians, *Palaeoniscum* had a shark-like tail with a larger upper lobe. It was probably a fast-swimming predator.

Primitive, non-teleost actinopterygians held sway during the Mesozoic. One noteworthy genus is *Dapedium*, collected from the Lower Jurassic of Lyme Regis in Dorset by, among others, famed fossil collector Mary Anning (1799–1847). This deep-bodied fish has a body covered with ganoid scales. These are thick and shiny thanks to their outer enamel-like layers, and articulate with one another, enclosing the animal in the equivalent of a suit of armour. The strong peg-like teeth of *Dapedium* point to a shell-crushing capability, so the fish probably feasted on a diet of molluscs. While the ganoid scales are the outstanding feature of *Dapedium*, the related genus *Lepidotes* is known mostly from its teeth. *Lepidotes* teeth are stud-like, typically brown or black in colour with a shiny surface. They are most often found singly but are occasionally recovered set in a cluster on a palate. Stomach contents from one example reveal that it ate crustaceans. The striking teeth of *Lepidotes* were once known as toadstones in the belief that they are the mythical jewels present in the heads of toads. This idea, coupled with an attractive appearance, led to their use in medieval jewellery rings as an alternative to conventional gemstones.

LEFT Palate of stud-like *Lepidotes* teeth from the Early Cretaceous rocks of Hastings in East Sussex, England measuring 8 cm (3 in) across.

BELOW *Dapedium politum* is a deep-bodied, ray-finned fish especially well known from the Early Jurassic rocks of Lyme Regis in Dorset, England. This fossil is almost 35 cm (13¾ in) in length.

Leedsichthys is a huge Jurassic fish believed to be the largest teleost ever to have lived. It was named in honour of Alfred Leeds (1847–1917), a farmer and fossil enthusiast who collected bones of this giant close to his home near Peterborough, England during the late nineteenth century. Among his finds were tail bones enabling a complete tail 3 m (9 ft) long to be reconstructed. Unfortunately, the skeleton of *Leedsichthys* is patchily preserved and the size of the living fish can only be estimated but it may have reached 12–17 m (39-56 ft) in length. The lack of teeth and the structure of the gill basket strongly suggest that *Leedsichthys* fed on plankton, rather like the similar-sized whale shark (*Rhincodon*) of today.

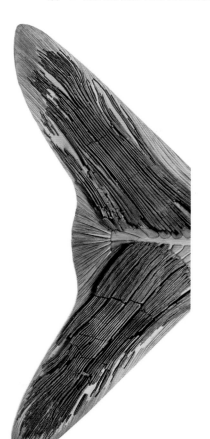

Modern coral reefs are vibrant ecosystems inhabited by diverse and colourful teleost fish. Nevertheless, ancient coral reefs are seldom rich in fossil fish of any kind because carcasses were quickly scavenged. An exception are the reef fishes from Monte Bolca close to Verona in Italy, an outstanding fauna that has been known to naturalists for almost 500 years. A staggering 200 different teleost species have been recorded at Monte Bolca, including the oldest known relatives of groups of fishes found in modern coral reefs. Herring, boxfish, batfish, tuna, barracuda and sea bass are among the fishes present in this 50-million-year-old deposit. The fishes occur in fine-grained limestones believed to have been deposited close to the reefs they inhabited. Low oxygen conditions on the seafloor plus the development of biofilms of microbial organisms contributed to their exquisite preservation.

LEFT Measuring 3 m (118 ft) from tip to tip, this is a reconstructed tail of the huge teleost fish *Leedsichthys problematicus* from the Jurassic Oxford Clay of Peterborough in Cambridgeshire, England.

BELOW *Seriola prisca* is an amberjack fish from the famous Monte Bolca deposit of Late Eocene age near Verona in Italy. This particular fossil is 18 cm (7 in) long.

Lobe-finned bony fishes

A second group of bony fishes are the sarcopterygians or lobe-finned fishes. Although sarcopterygians are represented today by just six species of lungfish and two coelacanths, they are important as fossils, not least for their bearing on the origin of land-dwelling tetrapods. Sarcopterygian fins may contain bones, along with muscles to control their movement, including equivalents of the humerus and femur found in tetrapod forelimbs and hindlimbs, respectively.

Coelacanths were thought to have become extinct in the Cretaceous period until late 1938 when a living coelacanth was caught off the coast of East London in South Africa. Described the following year as *Latimeria chalumnae*, it was hailed by scientists and in the press as a 'living fossil'. This sarcopterygian fish is rare, inhabits deep water and can grow up to 2 m (6½ ft) in length. Fossil coelacanths range back to the Devonian with more than 100 species recognized, most having a distinctive three-lobed tail, the middle lobe extending slightly beyond the lower and upper lobes.

Lungfish are more closely related than coelacanths to tetrapods. As the name implies, these fishes can breathe air using their functional lungs to supplement oxygen obtained by the gills; the equivalent structures in modern coelacanths are filled with fatty tissues. Fossils of lungfish or their near relatives date back to the Devonian when they reached maximum diversity before declining from the Carboniferous onwards. Reduction in the bony component of the skeletons of later lungfish at least in part explains why their fossil remains are relatively uncommon. Perhaps the most numerous fossils of lungfish are the toothplates of the Triassic–Eocene genus *Ceratodus*, which are very similar to those of the living Australian lungfish *Neoceratodus*.

LEFT This toothplate of the Triassic lungfish *Ceratodus palmatus* from Ludwigsburg in Germany measures 5 cm (2 in) in width.

TOP Oblique view of a cast of the type specimen of *Tiktaalik roseae* from the Early Devonian of Ellesmere Island, Canada. Although incomplete, this fossil exceeds 40 cm (15¾ in) in length.

ABOVE A large tooth, 13 cm (5 in) long, of the 'tetrapodomorph' fish *Rhizodus hibberti* from the Gilmerton Ironstone, an Early Carboniferous rock outcropping near Edinburgh, Scotland.

Considerable recent research has been undertaken on 'tetrapodomorphs', fossil fishes more closely related to tetrapods than to lungfish. The teeth of a Carboniferous tetrapodomorph called *Rhizodus hibberti* are quite common fossils. They can be more than 13 cm (5 in) long and are pointed and gently curved, with a series of longitudinal ridges and grooves near the base.

Tetrapodomorphs are fossils belonging to the group containing the ancestors of the first land-dwelling tetrapods. It is impossible to know for certain whether any are on the direct line to tetrapods rather than side branches. However, different tetrapodomorph genera do serve to elucidate the successive evolutionary changes on the line to amphibians. A fish low down on this line is *Eusthenopteron*, found at the same Devonian locality of Miguasha as the placoderm *Bothriolepis* (p. 150). Although *Eusthenopteron* was probably fully aquatic, the bones of the skull and the presence of an internal nostril recall tetrapods. Closer to the origin of tetrapods is *Tiktaalik*. Fossils of this fish were discovered in 2004 on Ellesmere Island in Arctic Canada. Almost crocodile-like in overall shape, *Tiktaalik* probably grew up to 2.5 m (8 ft) in length, much larger than *Eusthenopteron*, and had a flattened skull quite unlike that of most fishes, a body covered with bony scales, and an elbow joint but pectoral fins fringed with fin rays rather than the digits found in the forelimbs of tetrapods.

Tiktaalik is believed to have lived in shallow streams, occasionally emerging out of the water supported by its strong pectoral limbs. But why might *Tiktaalik* have left the water? One possibility is to move between drying bodies of water. Another is to bask in the sun, as do some crocodiles, the heat speeding up digestion after feeding. For whatever reason, venturing out of water was made easier by the fact that *Tiktaalik* and other tetrapodomorphs had lungs in addition to gills. It is not thought that the lungs evolved with the specific purpose of colonizing the land. Instead, they probably developed as an adaptation to living in waters deficient in oxygen, the fish taking gulps of air to supplement the oxygen acquired through the gills.

Primitive amphibians

Two fossils collected almost a hundred years ago in Greenland – *Acanthostega* and *Ichthyostega* – exhibit a combination of fish-like and tetrapod-like characters. Notably, these Late Devonian amphibians have limbs with bony digits (that is, fingers and toes) but a distinctly fish-like tail. *Acanthostega* was about 60 cm (2 ft) long, but *Ichthyostega* was at least twice that size. Both were better adapted to life on land than tetrapodomorphs such as *Tiktaalik*, especially *Ichthyostega*, which had a strong rib cage, although *Acanthostega* probably spent much of its time swimming through shallow-water swamps. Surprisingly, *Acanthostega* had eight fingers, and *Ichthyostega* seven. Tetrapod fossils with the five digits that are the norm today are not known before the latest Early Carboniferous.

Most Palaeozoic amphibians, like living amphibians, were relatively small but these have attracted less interest than a few gigantic genera. Temnospondyls are a group of Carboniferous–Cretaceous amphibians numbering over 150 genera. The largest

LEFT Model of *Acanthostega*, a primitive Late Devonian amphibian described from fossils collected in East Greenland.

ABOVE Viewed from above, the eye sockets are clear to see in the skull of the large temnospondyl amphibian *Eryops megalocephalus* from the Permian of Texas, USA. This fossil measures 30 cm (11¾ in) across.

are found in the Triassic and grew more than 5 m (16½ ft) in length, but the best known is the Permian genus *Eryops* which reached 2.5 m (8 ft) in length. This heavily built amphibian with a massive head was adapted for moving on the land, yet its teeth and their arrangement suggest that *Eryops* fed mostly on fishes in ponds and rivers. It seems possible that it had a lifestyle broadly similar to a crocodile.

The skull of another Palaeozoic amphibian – *Diplocaulus* – is extraordinary. This Carboniferous–Permian animal has a skull shaped like a boomerang or some might say Napoleon's hat, with a pair of long, backward-facing protrusions. Modelling shows that this unusual shape may have served as a means of generating lift during swimming movements. With its long, laterally compressed tail and feeble limbs, *Diplocaulus* was probably fully aquatic.

Lissamphibians

More than 8,000 species of amphibians are living today, far too many of which are threatened by extinction. Modern amphibians belong to a subgroup called the lissamphibians with three orders: frogs and toads, salamanders and newts, and the snake-like caecilians. Lissamphibians had their evolutionary roots in the temnospondyls, and all three orders are believed to have first diversified in the Triassic or Early Jurassic. Despite their high diversity in the modern biota, fossils of lissamphibians are comparatively rare, and fossils of caecilians remained unknown until the 1970s. The small average size of lissamphibians means that they have thin bones that are not readily fossilized. This said, some remarkable examples of fossil frogs have been discovered.

Miocene lake deposits in Libros, northeast Spain consist of deep-water, laminated mudstones containing intact fossils of the frog *Pelophylax pueyoi*. Not only are the bones of the frogs preserved but so too are many of the soft parts, including the stomach and nerves, with the dark outline of the flattened bodies clearly defined against the mudstone matrix. As is so often the case, replacement of tissues by phosphate minerals during decay is responsible for the exceptional preservation of the Libros frogs.

ABOVE *Diplocaulus* is a Late Palaeozoic amphibian with a skull of unmistakeable shape measuring 30 cm (11¾ in) or more across.

LEFT Missing only the end of its tail, this fossil tadpole, probably from the Cenozoic of Turkey, is over 7 cm (2¾ in) in length. Traces of the eyes are clearly visible.

As every schoolchild is taught, frogs are preceded by tadpoles, an aquatic larval stage that eventually metamorphoses into an adult frog. Remarkably, fossil tadpoles have been discovered in almost 50 ancient lake deposits ranging in age from Early Cretaceous to Miocene. Although a few of these fossil tadpoles are represented only by isolated bones, most consist of articulated skeletons (or moulds of the skeleton), and some exhibit soft part preservation with the distinctive outline of the body clearly evident. The fossil record certainly throws up a few surprises when it comes to what can, on occasions, be preserved!

9 Reptiles and birds: the rise of the archosaurs

The evolution of a membrane called an amnion was a key innovation in vertebrate evolution. The amnion encloses a fluid in which the embryo is contained, lessening the animal's dependence on watery habitats for reproduction and opening the way for a more complete terrestrial lifestyle than is possible in fishes and amphibians. Vertebrates possessing an amnion are known as amniotes. They include reptiles and birds, which are the subject of this chapter, and also mammals, which are covered in Chapter 10.

The two characteristics shared by present-day reptiles – lizards and snakes, turtles and tortoises, crocodiles and alligators – that most easily spring to mind are their scaly skins and cold-bloodedness. Fossil reptiles seldom preserve scales, apart from occasional impressions in the rock and, in any case, scales are lacking in some extinct species as described below. Cold-bloodedness can sometimes be inferred by estimating growth rates using growth rings in fossil bones: slow growth points to ectothermy (cold-bloodedness) while rapid growth suggests endothermy (hot-bloodedness). Using this method, it turns out that quite a lot of extinct reptiles were endothermic.

Defining a reptile is complicated further when phylogeny is considered: reptiles are not a 'natural group' because birds evolved from within them. The ancestry of birds can be traced to a particular group of dinosaurs, and the term 'non-avian dinosaur' is nowadays often used to refer to the dinosaurs of popular understanding and excludes birds. Reptiles plus birds are referred to by some scientists as sauropsids, a name which is unfortunately too easily confused with the dinosaur group called sauropods.

Amniotes show three patterns of openings in the sides of their skulls. All have a nasal opening and an orbit for the eye on each side of the skull, anapsids having only these. However, synapsids have an additional opening at the back of the skull, whereas diapsids have two such openings here. The additional openings make the skull lighter and allow for the passage of muscles. Among extant forms, chelonians (tortoises, turtles and terrapins) are 'anapsids' (although it is now believed that they started off

OPPOSITE Mounted specimen of the huge sauropod dinosaur *Patagotitan mayorum*, which was first discovered in 2010 in Early Cretaceous rocks near La Flecha in Argentina.

as diapsids but the two openings in the skull were secondarily closed), mammals are synapsids, and crocodilians (crocodiles and alligators), rhynchocephalians (tuatara), lepidosaurs (lizards and snakes) and birds are diapsids. Extinct anapsids include various early reptiles, whereas pelycosaurs and the so-called mammal-like reptiles (see Chapter 10) are synapsids. Diapsids include terrestrial dinosaurs, flying pterosaurs, and swimming ichthyosaurs, plesiosaurs and mosasaurs.

Early reptiles

The fossil record of reptiles begins in the Carboniferous. Reptiles apparently emerged from temnospondyl amphibians probably similar to *Westlothiana lizziae*, a 20-cm (8-in) lizard-like species discovered in 1984 at East Kirkton Quarry in Scotland. Slightly younger rocks in Nova Scotia, Canada, contain footprints with scale impressions that are more unequivocally reptilian.

Large-sized reptiles proliferated in the Permian, not least a group called pelycosaurs, which reached up to 4 m (13 ft) in length and are sometimes mistaken for dinosaurs. The best known pelycosaurs are the sail-backed genera *Dimetrodon* and *Edaphosaurus*. The former has sharp teeth indicating that it was a carnivore, whereas *Edaphosaurus* is equipped with batteries of peg-like teeth suggesting a herbivorous diet. But the most

BELOW The sail-backed pelycosaur *Dimetrodon grandis* from the Early Permian of Texas, USA. Displayed in the National Museum of Natural History in Washington DC, this mounted specimen measures 3.1 m (10 ft) in length.

striking features of these two genera are the immense prolongations on the vertebrae that supported the flesh of the large dorsal sail. The function of the pelycosaur sail is uncertain, some considering it to have been used for thermoregulation, others in courtship displays. Pelycosaurs are synapsids, more closely related to mammals than they are to 'classical' reptiles. Further along the mammal line are the therapsids discussed in the next chapter.

Reptiles with shells

Chelonians – turtles, tortoises and terrapins – need little introduction. A modest 350 species or so of these reptiles are living today, some species popular as pets but far too many teetering on the brink of extinction. The defining feature of chelonians is the shell that encloses the body in a protective case. The shell has two main parts – a carapace on the upper side and a plastron on the lower side. These are constructed from a complex of modified ribs and other skeletal bones as well as dermal bones and keratinous 'scutes'. The latter can be shed periodically in aquatic turtles and terrapins.

Because of their shells, chelonians have a better fossil record than most reptiles. The oldest clear examples occur in Triassic rocks, notably the 220-million-year-old *Odontochelys* from China. Whereas teeth are missing in modern chelonians, this Triassic chelonian has small teeth. The shell lacks a carapace and consists of a plastron only, indicating that this lower component of the shell evolved first. *Odontochelys* was aquatic, inhabiting coastal environments. This suggests that the first chelonians lived in water, with land-dwelling tortoises evolving later. The skull of the slightly younger Triassic turtle *Proganochelys* shows some primitive features that are lost in later chelonians.

By the Cretaceous, three main groups of chelonians had appeared: the extinct protostegids, the extant hard-shelled cryptodires and the leather-backed dermochelyids. *Archelon* is a remarkable protostegid discovered in the Late Cretaceous Pierre Shale of South Dakota, USA. Reaching up to 4.6 m (15 ft) in length, this is the largest chelonian known, either fossil or living. The upper jaw has a beak-like hook suggesting that, unlike modern turtles, *Archelon* was a predator, perhaps using the hook to puncture mollusc shells.

ABOVE Measuring 5 cm (2 in) in length, a skull of the early turtle *Proganochelys quenstedti* from the Triassic of Wurttenberg, Germany.

RIGHT Well-preserved carapace of the tortoise *Stylemys nebrascensis* from Nebraska, USA measuring about 20 cm (7¾ in) long.

The evolutionary history of chelonians reveals numerous transitions between marine, freshwater and terrestrial lifestyles. An example of a terrestrial genus, in other words a tortoise, is *Stylemys* found in the Eocene and Oligocene. Based on the structure of the limbs, *Stylemys* lacked the ability to burrow, unlike many present-day tortoises.

Lizards and the like

The New Zealand tuatara *Sphenodon punctatus* is a remarkable survivor, the only living species of a group called the rhynchocephalians. Although *Sphenodon* is at first glance easily mistaken for a true lizard, it differs among other features in having a second, inner tooth row. Rhynchocephalians were much more abundant and varied in the geological past. The oldest fossils come from Middle Triassic rocks and are about 240 million years old. Slightly younger examples have been collected from sediments filling fissures and caves in limestones. Although they prospered during the Jurassic, with one group – the pleurosaurids – evolving elongate bodies adapted for swimming, an inexorable decline in their fortunes had set in by the Early Cretaceous.

Rhynchocephalians, together with lizards and snakes, constitute a group called lepidosaurs. But whereas rhynchocephalians today are hanging on by the narrowest of threads, over 7,000 lizard species and about 3,500 snake species are living today. The fossil record of lizards is patchy and relatively few species are represented by all the bones of the body, limb bones often being unknown. The oldest recorded fossil lizard was first described in 2022. This is *Cryptovaranoides microlanius*, which, like the rhynchocephalian *Clevosaurus*, comes from Late Triassic fissure fillings in the southwest of Britain. *Cryptovaranoides* resembled some modern lizards and may have grown to a length of about 25 cm (10 in). Most fossil lizards are small, including some tiny enough to be encapsulated in small pieces of amber. A notable exception is the gigantic Eocene lizard *Barbaturex morrisoni* from Myanmar, named for Jim Morrison, singer with the rock band The Doors. Judging from the only bones known, which are jaw fragments, this herbivorous iguanid may have reached almost 1.8 m (6 ft) in length.

Bones and teeth belonging to the so-called 'Monster of Maastricht' were discovered during the mid-eighteenth century in a mine extracting limestone from close to Maastricht in the southeast of the Netherlands. This fossil reptile was subsequently recognized as a mosasaur, a group of large marine lizards found only in Late Cretaceous rocks. Ranging from 0.3 to 18 m (1-59 ft) in length, mosasaurs were ferocious predators with sharply pointed teeth. Puncture marks found in some ammonite shells match the pattern of the teeth in mosasaur jaws, suggesting that ammonites were among their prey.

A remarkable feature of lizard evolution has been the multiple instances of reduction and eventual loss of the limbs, generally in association with a burrowing lifestyle. One such change resulted in the evolution of snakes. While some advanced snakes are able to kill their prey using constriction and to open their jaws enough to swallow prey wider than their own bodies, the first snakes may have employed more conventional means to hunt at night. The oldest well-preserved fossil snakes date from the Late Cretaceous, although fragmentary Jurassic fossils suggest the existence of earlier species. The largest known snake is a fossil species called *Titanoboa cerrejonensis* found

ABOVE Skull of the Triassic rhynchocephalian *Clevosaurus*, a relative of the modern tuatara of New Zealand. This fossil was collected in the Mendip Hills of southwest England and measures about 3.5 cm (1¼ in) across.

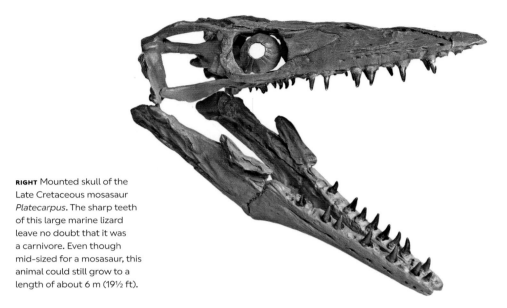

RIGHT Mounted skull of the Late Cretaceous mosasaur *Platecarpus*. The sharp teeth of this large marine lizard leave no doubt that it was a carnivore. Even though mid-sized for a mosasaur, this animal could still grow to a length of about 6 m (19½ ft).

in the Paleocene rocks of Colombia. Its huge vertebrae allow the length of this snake to be estimated as at least 13 m (42 ft), far larger than today's anacondas. A more modest-sized snake represented by its vertebrae is *Paleryx rhombifer*, first described in 1850 from the Late Eocene of Hordle Cliff in Hampshire, England, the earliest description of a fossil constrictor snake.

Crocodiles and their kin

Today's crocodiles and alligators, along with a plethora of related extinct reptiles, constitute a group known as the crocodylomorphs. Crocodylomorphs in turn are part of a larger group – archosaurs – embracing dinosaurs and birds among others, that diverged at least 255 million years ago from the reptile group containing lizards and their close relatives. At the base of the crocodylomorph evolutionary tree are

BELOW An 80 cm (31½ in) long skull of the crocodile *Plagiophthalmosuchus gracilirostris* collected from Early Jurassic rocks at Whitby on the coast of North Yorkshire, England.

the Late Triassic–Middle Jurassic sphenosuchians. With longer legs but a less elongate skull than later crocodylomorphs, many sphenosuchians were fully terrestrial animals lacking the squat posture of modern crocodiles.

ABOVE Dermal scute of a Jurassic crocodile from Peterborough, Cambridgeshire, England. About 4 cm (1½ in) across.

The fossils of crocodylomorphs most likely to be found are teeth and dermal scutes, bony plates just beneath the skin typically with distinctive pitted surfaces. In some modern crocodiles, scutes are small or absent, as is the case in the saltwater crocodile *Crocodylus porosus*, which is unique among the 24 species of crocodylomorphs living today in inhabiting the sea rather than freshwaters. However, marine crocodylomorphs were more diverse in the geological past, notably those which lived alongside ichthyosaurs and plesiosaurs in Mesozoic seas. The Early Jurassic thalattosuchian *Plagiophthalmosuchus gracilirostris* had a long snout to match the length of its genus name, and as narrow as implied by its species name, *gracilirostris*, which means delicate snout. First found in Early Jurassic rocks of Whitby, Yorkshire, the slender teeth of this species are consistent with a diet of fishes.

Visitors to the fossil gallery of the Muséum national d'Histoire naturelle in Paris cannot help but be impressed by the mounted skeleton of one of the largest known crocodiles, *Sarcosuchus imperator*, nicknamed 'Supercroc'. Found in Early Cretaceous rocks in Algeria and Niger, this crocodile may have grown to over 9 m (29½ ft) in length and is believed to have fed at least in part on dinosaurs, dragging these large animals from riverbanks into the water.

Swimming reptiles: ichthyosaurs and plesiosaurs

Two types of large marine reptiles – ichthyosaurs and plesiosaurs – evolved from terrestrial ancestors to become apex predators in Jurassic and Cretaceous seas. Ichthyosaurs had streamlined bodies resembling dolphins, but with vertical rather than horizontal tails and a pair of rear fins. Plesiosaurs, on the other hand, had barrel-shaped bodies and four flipper-like limbs. Some had long necks and small heads, while others had short necks and large heads; they are perhaps best known in the guise of the mythical Loch Ness Monster. Both groups are moderately common as fossils because they inhabited nearshore waters where sunken carcasses stood a reasonable chance of being buried by sediment and entering the fossil record. British Jurassic rocks, especially along the Dorset and North Yorkshire coasts where erosion is constantly uncovering new fossils, have been revealing ichthyosaurs and plesiosaurs since the early eighteenth century, not just isolated bones but also fully articulated skeletons.

Colloquially known as 'sea dragons', more than 100 genera of ichthyosaurs have been described to date. They range from the Early Triassic to the mid-Cretaceous, a duration of some 160 million years. Most species measure a few metres in length, but the smallest was just 0.3 m (1 ft) while the largest may have been as much as 25 m (82 ft) long.

While ichthyosaur vertebrae and rib bones are not uncommon fossils, the rarer articulated skeletons – especially those showing soft part preservation – have made possible a detailed understanding of the morphology and ecology of these animals. Fossils showing outlines of the complete body in the form of dark-coloured films surrounding the skeleton have been known for almost 200 years. These show that the animals had a dorsal fin, like that of a shark but fleshy, and a tail comprising an upper lobe without bones and a lower lobe containing the downwardly deflected backbone. More recently, remnants of scaleless skin underlain by an insulating layer of blubber, suggesting a warm-blooded physiology, have been described from an exceptional fossil of the Early Jurassic ichthyosaur *Stenopterygius*.

Stomach contents indicate that ichthyosaurs fed mainly on cephalopods and fishes. A striking feature of ichthyosaur heads is a ring of bones which would have

ABOVE Skull of the Early Jurassic ichthyosaur *Ichthyosaurus communis*. Note the sclerotic ring within the eye socket.

surrounded the eyes. The size of these so-called sclerotic rings can be used to estimate eye size. This shows that ichthyosaurs had large eyes relative to the overall size of their bodies, probably allowing them to have hunted in deeper water where light levels were low. In addition, they would have bestowed the visual acuity needed to capture cephalopods, despite the ability of these animals to release ink into the water to mask their presence.

The fact that ichthyosaurs gave birth to live young has been known since 1845 when Joseph Chaning Pearce (1811–1847) discovered a small embryo in the position of the birth canal of an adult individual. Several similar examples have since been described.

So many pregnant specimens of *Stenopterygius* have been found in the Early Jurassic Posidonia Shale of Holzmaden in southern Germany that it is possible this was a site of congregation for females about to give birth.

Over 120 genera of plesiosaurs have been described, ranging in length from less than 2 m (6½ ft) to more than 14 m (46 ft). The oldest

LEFT *Ichthyosaurus communis* skeleton from the Early Jurassic of Street in Somerset, England. This individual measures about 2.4 m (8 ft) in length.

ABOVE Plaster cast of the plesiosaur *Eretmosaurus rugosus*. The head, which would have been on the far right, is missing. The original specimen was excavated from Early Jurassic rocks at Granby in Nottinghamshire, England.

known plesiosaur comes from the Late Triassic, while the last plesiosaur perishing during the end-Cretaceous mass extinction. Although predominantly marine animals, some plesiosaurs were able to colonize brackish and even freshwater environments, including rivers.

Two bodily shapes occur among plesiosaurs: the graceful plesiosauromorphs with small heads and long necks containing up to 76 vertebrae and measuring as much as 9 m (29½ ft) in length, and the stockier pliosauromorphs. The latter have larger heads and shorter necks, with as few as 35 vertebrae. The swimming technique of plesiosaurs was unlike that of any other vertebrates. It involved 'flying' through the water driven by all four limbs, which are modified into equal-sized pointed flippers. The short tails of

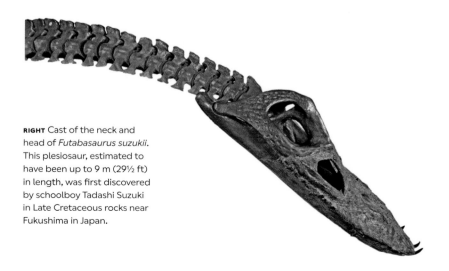

RIGHT Cast of the neck and head of *Futabasaurus suzukii*. This plesiosaur, estimated to have been up to 9 m (29½ ft) in length, was first discovered by schoolboy Tadashi Suzuki in Late Cretaceous rocks near Fukushima in Japan.

plesiosaurs were used for balance and as rudders rather than for providing propulsion.

The conical teeth have very sharp cutting edges in some pliosauromorphs, useful for dealing with large prey such as ichthyosaurs, which they pursued at speed. In contrast, plesiosauromorphs are thought to have been slower moving ambush predators that hunted among schools of fish and cephalopods. It has long been known that plesiosaurs swallowed stones that accumulated in their stomachs as gastroliths. Other animals, living (e.g. crocodiles and birds) and extinct (e.g. dinosaurs), also possess gastroliths. It is likely that they play a role in digestion, aiding the breaking down of food. No fewer than 289 gastroliths were found associated with a plesiosaur skeleton excavated from the Late Cretaceous of southern Utah, USA.

ABOVE Tooth of the Jurassic pliosaur *Liopleurodon ferox*, which can be up to 13 cm (5 in) long.

It was formerly believed by many scientists that plesiosaurs laid eggs, hauling themselves ashore to bury their eggs in the sand, as turtles do today. However, in 1987 an adult of the plesiosaur *Polycotylus* was discovered in Kansas, USA with a foetus preserved within its skeleton. This proved that plesiosaurs, like ichthyosaurs and mosasaurs, gave birth to live young. In addition to being viviparous, the microstructure of plesiosaur bones suggests that they were warm-blooded.

Flying reptiles: pterosaurs

A popular misconception is that pterosaurs – 'pterodactyls' in colloquial parlance – are flying dinosaurs. Although this is incorrect, pterosaurs are very closely related to dinosaurs: both are archosaurs, they share a common ancestor and lived at approximately the same time. Fewer than 200 species of pterosaurs have been described. Most articulated fossil pterosaurs are preserved flattened over bedding planes. The bones themselves are typically delicate and contain air sacs that formed part of the respiratory system, as in birds and non-avian dinosaurs.

Pterosaurs have sharply pointed teeth inclined forwards in some species, with those at the front being longer to grab prey. Although often portrayed as fish-eaters, it is believed that some pterosaurs may have fed on insects, terrestrial vertebrates or even plant leaves and fruits. Striking crests of various shapes characterize many pterosaurs. Crests range from low ridges extending along the jaws (e.g. *Anhanguera*) to extravagant triangular structures projecting from the rear of the skull (e.g. *Pteranodon*). Many crests are constructed of bone; others comprise soft tissues and are only apparent in exceptionally preserved fossils. Although crests probably helped to stabilize the orientation of the head when flying or fishing, there is some evidence that the males and females of genera such as *Darwinopterus* had crests of different shapes, implying a role in sexual selection.

Pterosaurs were the first vertebrates to evolve powered flight. They made this breakthrough some 75 million years before birds and 175 million years before bats. The independent acquisition of flight in these three groups is a textbook case of convergent evolution. Pterosaurs, birds and bats all have wings but of radically different structure. The wings of pterosaurs are supported along their leading edges by the greatly elongated fourth digit of the forelimbs, and are also normally attached to the hind limbs. Birds, however, have wings formed by a string of arm bones (humerus, radius, ulna, metacarpus and to a lesser extent three of the digits). In contrast, bat wings comprise arm bones plus elongated digits two to five, with the wing also attached to the hindlimb, as in pterosaurs. The wings themselves are constructed differently in the three groups: pterosaur wings comprise a membrane reinforced by supporting fibres (actinofibrils) and containing a thin layer of muscles; bird wings consist of numerous separate flight feathers; and bat wings are made of stretched skin.

BELOW This jumble of bones is a fossil of the pterosaur *Dimorphodon macronyx* discovered by Mary Anning in the Early Jurassic rocks of Lyme Regis, Dorset, England. The toothed skull is visible on the left.

LEFT Computed tomographic (CT) image of the skull of the Cretaceous pterosaur *Anhanguera*. This fish-eating flying reptile had a wingspan exceeding 4 m (13 ft).

Although pterosaurs lacked true feathers, their bodies were covered by hair-like filaments called 'pycnofibres', each about 5–7 mm long. The coating of pycnofibres insulated the body against the cold, evidence that pterosaurs were warm-blooded animals. Pterosaur wingspans ranged from 40 cm (16 in) to an incredible 10 m (33 ft), almost three times that of the wandering albatross, which has the largest wingspan of living birds. Modelling has shown pterosaurs to have been adept in flight, despite weighing as much as an estimated 250 kg (550 lb). Trackways left by pterosaurs show that they were quadrupedal walkers, but the manner in which they took to the air is still a matter of debate among palaeontologists.

New discoveries of pterosaurs are being made at an increasing rate, particularly in Brazil and China. As recently as 2004, eggs were first reported in pterosaurs. These had soft shells, a trait shared with the eggs of many other reptiles.

Dinosaurs

Scarcely a day goes by without a dinosaur story in the media. The public – children especially – have an insatiable thirst for these monsters of the Mesozoic. The number of palaeontologists studying dinosaurs has ballooned, with a recent online article published by a highly respected news organization even going as far as defining a palaeontologist as a scientist who studies dinosaurs! With all this public and scientific interest, it is not surprising to learn that a new species of dinosaur is currently being recognized almost every week, adding to the more than 1,000 species already described. Nevertheless, unless prospecting in special places like the Badlands of the western USA, you are far less likely to come across a dinosaur bone than the other sorts of fossils that attract scant attention. Countless books have been published on dinosaurs, and these should be consulted for fuller accounts than is possible here.

Almost 40 years before he became the first 'chief superintendent' of the newly founded British Museum (Natural History) in 1881, the leading anatomist Richard Owen (1804–1892) coined the name Dinosauria, meaning terrible reptile. Owen's

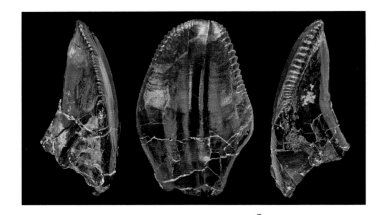

RIGHT Three views of a tooth of *Iguanodon* from the Early Cretaceous of East Sussex, England, one of the first dinosaurs to have been discovered. The chisel-like tooth is about 4 cm (1½ in) wide.

BELOW Cast of the jaw of *Tyrannosaurus rex* measuring 1.2 m (4 ft) long. This dinosaur was a formidable predator living in North America towards the end of the Cretaceous period.

new group of reptiles was founded on three genera – *Megalosaurus*, *Hylaeosaurus* and *Iguanodon* – described earlier in the nineteenth century based on bones discovered in southern England, the first from Middle Jurassic rocks in Oxfordshire and the other two from Early Cretaceous rocks in Sussex. The remains of dinosaurs have since been found on all continents of the Earth, with hotspots for dinosaur discoveries being western North America, Argentina, Mongolia and China. Dinosaurs were distributed very widely across the Mesozoic globe. Some dinosaurs even inhabited the Arctic where, although polar ice-caps were lacking, they faced cold temperatures and winter days of continuous darkness. Remarkably, during the Jurassic and Cretaceous every known animal over 1.5 m (5 ft) in size that walked on land was a dinosaur.

Two main groups of dinosaurs are recognized: ornithischians and saurischians. Ornithischians include such iconic genera as *Stegosaurus*, *Iguanodon* and *Triceratops* as well as the 'duck-billed' hadrosaurs of the Late Cretaceous. Some ornithischians walked on all four limbs, others were bipedal. Most were herbivores, although it is possible that some of the first ornithischians from the Early Jurassic were omnivores. The saurischians comprise two groups, sauropods and theropods. Sauropods are the gigantic, long-necked, quadrupedal herbivores, typically found in Jurassic deposits, such

LEFT Restoration of the small, feathered dinosaur *Microraptor* from the Early Cretaceous Jiufotang Formation of Liaoning in China. Although shown with bright plumage, more recent research indicates that the feathers may have been iridescent black in colour.

as *Diplodocus*, *Brontosaurus* and *Brachiosaurus*. Some of the first sauropodomorphs, like *Buriolestes* from the Late Triassic, however, may have subsisted on mixed diets of leaves and small animals. The other group of saurischian dinosaurs are the theropods, which includes the predatory dinosaurs, notably *Tyrannosaurus* and *Velociraptor*, as well as birds.

There can be no doubting the dominant role played by dinosaurs in terrestrial ecosystems during the Jurassic and Cretaceous. From agile predators with powerful claws and sharp teeth, to lumbering long-necked giants and squat heavily armoured herbivores, dinosaurs differed enormously in morphology, behaviour and ecology. Particularly striking is the remarkable range in body sizes among non-avian dinosaurs. The smallest known non-avian dinosaur is an Early Cretaceous species of *Microraptor* from China. This theropod measured about 77 cm (27½ in) in length and is estimated to have weighed between 0.5 and 1.4 kg (1–3 lbs). Aside from its small size, *Microraptor* is also notable in having feathers on both its fore- and hind-limbs, probably used in gliding. At the other end of the size spectrum is the sauropod *Patagotitan mayorum*, 37 m (121 ft) long, 8.5 m (28 ft) high and weighing as much as 70,000 kg (77 tons), equivalent to roughly 100,000 individuals of *Microraptor*!

On average, a dinosaur skeleton comprises about 200 individual bones. Very few species are known from a complete set of bones. *Patagotitan mayorum*, for instance, is represented by 60% of the bones. As in many other sauropods, bones of the head have not been discovered and its morphology must be inferred from related species. Almost half of dinosaur genera were described based on just a single bone.

Aside from bones and teeth, dinosaurs have also left their mark in the fossil record as footprints and trackways (see Chapter 13) and eggs. Dinosaur eggs vary in shape from almost perfectly spherical to up to three times longer than wide, range in size from 10 to 30 cm (4–12 in) in diameter, and are often covered by small pores and bumps. Usually made of layers of calcite, dinosaur eggs were laid in clutches in nests.

RIGHT Cast of a nest of eggs thought to have been laid by the dinosaur *Protoceratops*. Discovered in Late Cretaceous rocks of the Gobi Desert, Mongolia, each egg is about 15 cm (6 in) long.

BELOW Humerus bone from the early dinosaur *Nyasasaurus parringtoni* found in the Middle Triassic rocks of Tanzania. The animal may have been 2–3 m (6½–9¾ ft) long.

Some eggs contain the bones of embryos, which can be studied in great detail using computed tomography scanning techniques.

An illuminating recent study on *Tyrannosaurus rex* sought to estimate the total number of individuals of this dinosaur that ever lived and to compare it with the number of fossils that have been collected. Using a formula predicting the population density of animals of given mass and knowing the geographical range of *T. rex*, it was calculated that about 20,000 individuals of *T. rex* would have been alive at any one time. As *T. rex* existed for about 2 million years, there would have been roughly 127,000 generations between its origin and final extinction, giving a total of 2.5 billion individuals of *T. rex* over its entire duration. A mere 32 adult individuals of *T. rex* have been collected as fossils – one in 80 million of those estimated to have lived – despite the substantially higher commercial value of *T. rex* skeletons than those of other dinosaurs, which makes them especially attractive to fossil hunters and dealers.

The geological history of dinosaurs can be traced back to the Triassic period. A candidate for the oldest known dinosaur is *Nyasasaurus* from the Middle Triassic of Tanzania, which is about 245 million years old. Unfortunately, this genus is known only from a few vertebrae and an arm bone, precluding an unequivocal placement among the dinosaurs. Early dinosaurs and their closest relatives seem to have lived in the shadow of larger animals belonging to other groups of reptiles such as crocodilian archosaurs. The mass extinction marking the end of the Triassic period 201 million years ago removed these larger animals, clearing the way for dinosaurs to diversify in the Early Jurassic. A rich panoply of dinosaur genera followed in the Jurassic and Cretaceous, with rapid evolutionary turnover, most genera probably surviving for less than 10 million years.

The final demise of the non-avian dinosaurs occurred during the end-Cretaceous mass extinction event (p. 26). The horn of a *Triceratops* was recently discovered in the Hell Creek Formation of Montana, USA just 12 cm below the extinction horizon, whereas bones belonging to other dinosaur genera (e.g. the hadrosaur *Edmontosaurus*, and *Tyrannosaurus*) have been found in the underlying, slightly older rocks in the same area.

Birds

With more than 10,000 living species, flamboyant plumage and fascinating behaviours, birds attract countless enthusiastic naturalists. When it comes to the fossil record, however, they offer much poorer pickings. The fossil record of birds is patchy, in part because of their fragile, lightweight bones and lack of teeth. In addition, the upland habitats favoured by many species are not well-represented in the rock record.

In the first edition of *On the Origin of Species* in 1859, Charles Darwin admitted that the absence of a missing link between the closely related reptiles and birds was a concern for his theory of evolution. Just 2 years later, the discovery of *Archaeopteryx* in the Late Jurassic Solnhofen Limestone of Bavaria provided Darwin with the missing link he required. *Archaeopteryx* is the oldest known bird fossil, and combines features associated with reptiles and those more typical of birds: it has teeth and a long bony tail like a reptile, but feathers and a wishbone like a bird. Only thirteen examples of this celebrated fossil have been found, plus some impressions of isolated feathers in the fine-grained limestone.

Although it used to be thought that the evolutionary radiation of birds was delayed until after the end-Cretaceous mass extinction, more than 120 species of birds have now been described from the Mesozoic and molecular evidence points to the presence of ancestors of many extant groups that remain undiscovered in the fossil record. The Chinese genus *Eoconfuciusornis* from the Early Cretaceous is the oldest known fossil bird with a beak – as opposed to teeth – appearing some 20 million years after *Archaeopteryx*. Remarkable preservation shows the plumage of males and females of *Eoconfuciusornis* to have differed, the males possessing a pair of elongated tail feathers that were lacking in the females. Another important group of Mesozoic birds are the hesperornithiformes. Numbering about 20 species, these birds from the Late Cretaceous have small teeth,

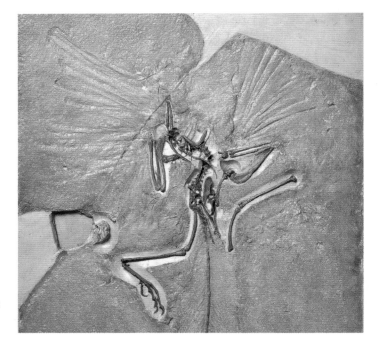

RIGHT The 'London specimen' of *Archaeopteryx lithographica*. Fossils of this primitive bird, which had a wingspan of about 60 cm (23½ in), come from the Lithographic Limestone of Solnhofen in Bavaria, Germany, and are Late Jurassic in age. Note the feather impressions.

long necks and streamlined bodies, the largest reaching 1.5 m (5 ft) long. The reduced forelimb and robust hindlimb bones show hesperornithiformes to have been flightless, foot-propelled diving birds like modern loons.

Ostriches, emus and other flightless birds classified as ratites constitute the bulk of one of the two major groups of birds, the palaeognaths. Far less diverse than the neognaths, to which over 99% of modern birds belong, in palaeognaths the bones forming the roof of the mouth are more reptile-like. Moa are especially notable among extinct palaeognaths as one of numerous examples of birds driven to extinction by human beings. Nine species of moa lived in New Zealand until quite recently, the largest more than 3 m (9 ft) in height. They are represented by bones, eggs, footprints (p. 247) and occasional mummified soft tissues. Most are subfossils, less than 10,000 years old. Although the oldest moa fossils come from the Miocene, it has been estimated that their ancestors colonized New Zealand 60 million years ago, evolving to become flightless in a land free from large predators. Sadly, moa became an all too easy source of meat for the first human colonizers of New Zealand about 750 years ago.

Phorusrhacids are an unrelated group of large flightless birds, aptly named 'terror birds' because they were carnivores, unlike the herbivorous moas. Ranging from the Eocene to the Pleistocene, most of the 18 recognized species of phorusrhacids have

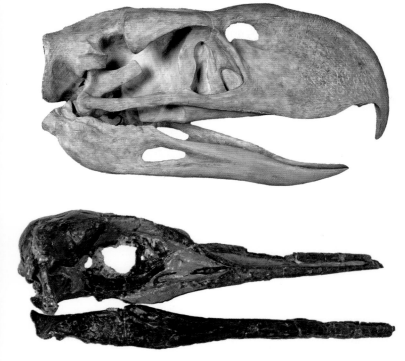

TOP LEFT The large, 59 cm (23 in) long, and powerful skull of *Patagornis marshi*, a flightless terror bird from the Miocene of Argentina.

LEFT A skull of *Prophaeton shrubsolei* from the Early Eocene of Sheppey, Kent, England is the type specimen of this species, which was a petrel-like bird. The fossil measures 12 cm (4¾ in) in length.

been found in South America, the largest standing 3 m (9 ft) high. It is believed that phorusrhacids were fast and highly manoeuvrable runners. At up to 46 cm (18 in) in length and usually hooked, their sharp beaks resembled those of eagles. These were formidable weapons, as were their claws.

Important finds of fossil birds have been made in the Eocene London Clay of England, which is also a famous source of fossil plants (see Chapter 11). Birds belonging to more than 25 different species were buried in the shallow sea of the London Clay in southern England. They include genera that either belong to, or are close relatives of, bird groups familiar today, e.g. loons, owls, parrots, storks, storm petrels, swifts and tropic birds. Perching birds – passerines – are believed to have originated at about the same time, but in the southern hemisphere.

An unmistakable southern hemisphere bird is the penguin. The oldest fossil penguins date back to the mid Paleocene, about 60 million years ago. *Kumimanu fordycei* is a recently discovered penguin from the late Paleocene of New Zealand remarkable for its enormous size. Estimated to have weighed up to 160 kg (353 lb), three times more than the modern Emperor Penguin, this giant was as tall as an adult human.

10 Our family and other mammals

There is an almost inevitable tendency to place mammals at the top of the evolutionary tree, with humans at the very summit. As all extant species have an equal claim to occupy positions at the tips of the branches, this view is seriously flawed. However, there is no denying the importance of contemporary mammals or their amazing adaptations to different environments and ways of living.

The earliest known mammals in the broadest sense come from Late Triassic rocks about 225 million years old. These few forerunners spawned the roughly 6,400 species of mammals that live today, a richness that can be attributed largely to the dramatic diversification of many groups of mammals following the K-Pg mass extinction event at the end of the Cretaceous, some 66 million years ago. Rodents are the single most diverse group of living mammals, accounting for approximately 40% of mammal diversity. They are followed, perhaps surprisingly, by bats.

Mammals are distinguished from other vertebrates by producing milk from modified sweat glands to suckle their young. This is even the case with the peculiar egg-laying monotreme mammals, the platypus and echidna. Not surprisingly, direct evidence of milk production is lacking in fossils but there are several features of the fossilized bones diagnostic of mammals. One of the most obvious is the lower jaw, which is constructed of a single bone in mammals but multiple bones in reptiles. Astonishingly, the two bones that form the articulation of the reptilian jaw are transformed in mammals into the incus and malleus bones of the middle ear.

The size of mammals varies massively between species, from the 200,000 kg (220 tons) of the blue whale down to the 2 g (0.002 kg; less than one-tenth of an ounce) of the pygmy shrew. While the bones of mammoths and other fossil species near the large end of the size spectrum attract the greatest public attention, the teeth of small, mouse-sized species dominate the mammal fossil record. Fortunately, fossil teeth, even when isolated, can reveal a lot about the diets of ancient mammals as well as allowing them to be assigned to a major group.

OPPOSITE Cast of the skull of a young male of *Paranthropus boisei*, a 1.8 million-year-old hominin from Oldupai Gorge in Tanzania.

There are three subclasses of living mammals: the egg-laying Monotremata, the pouched Marsupialia and the dominant Placentalia. Molecular data have led to massive changes in the high-level classification of placental mammals. Although not universally accepted, this method has recognized four superorders: Afrotheria (aardvarks, elephants and sea cows), Xenarthra (sloths, armadillos and anteaters), Laurasiatheria (carnivores, bats, whales and most hoofed mammals) and Euarchontoglires (primates, rodents and rabbits).

Therapsids and early mammals

Mammals belong to a wider group known as therapsids, which evolved from the pelycosaurs described in the previous chapter. Non-mammalian therapsids are often known as 'mammal-like reptiles', an apt name given their intermediate status. Like mammals, but unlike typical reptiles, these Permian to Early Cretaceous animals had teeth differentiated into incisors, sharp canines and molars. *Massetognathus* belongs to a therapsid group called the cynodonts. This Triassic animal was apparently covered in fur and had the stature of a medium-sized dog, although it was probably

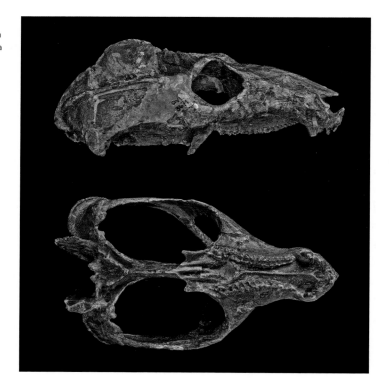

RIGHT This 18 cm (7 in) long skull is *Massetognathus* from the Middle Triassic of La Rioja Province in Argentina. The teeth of this cynodont are differentiated into several different types, generally a hallmark of mammals.

LEFT *Cynognathus crateronotus* was a carnivorous cynodont, as implied by the prominent canine teeth. The skull of this Middle Triassic of South Africa could grow to 40 cm (15¾ in) in length.

a herbivore that laid eggs. Another Triassic cynodont called *Cynognathus* was bigger and equipped with the large canine teeth of a carnivore.

Closer to true mammals are animals such as *Morganucodon* of Late Triassic to Middle Jurassic age. Fossils of this 'mammaliaform' genus were first found among sediments in South Wales, filling fissures eroded into Carboniferous limestone exposed on the Mesozoic land surface. Measuring about 10 cm (4 in) in length, *Morganucodon* was apparently an egg-laying insectivore and probably nocturnal in habit.

A prominent early group of mammals was the multituberculates in which the molar teeth have rows of small cusps. Ranging from the Middle Jurassic to the late Eocene, multituberculates were small insectivores or omnivores that lived in the shadow of the dinosaurs but diversified in the Late Cretaceous, at the same time as flowering plants became abundant. Multituberculates resembled rodents in some respects but had pointed teeth that could have been used for winkling seeds out of husks. Some scientists believe multituberculates to be closely related to monotremes, others to the group containing marsupial and placental mammals.

ABOVE Lower molar tooth of *Morganucodon watsoni*, a small cynodont from the Duchy Quarry in Glamorgan, Wales, probably of Early Jurassic age.

Monotremes and marsupials

Only five species of monotremes exist at the present day – the duck-billed platypus and four species of echidnas – all native to Australia or New Guinea. They are unique among modern mammals in laying eggs rather than giving birth to live young, and all are toothless. Just a handful of fossil genera are known, the oldest being *Teinolophos*. Fossils of this small animal, which weighed an estimated 40 g (1½ oz), have been found in Early Cretaceous rocks of Victoria, Australia. Unlike extant monotremes, *Teinolophos* possessed teeth including five sets of molars. *Monotrematum*, a Paleocene fossil found in Patagonia, is one of only two unequivocal monotremes from outside Australia or New Guinea. It is thought to have reached South America via Antarctica before the break-up of Gondwana.

Marsupials, mammals which carry their young in a pouch called a marsupium, are particularly associated with Australia, home to almost 70% of the 334 living species. Nevertheless, both fossil and molecular evidence suggests that marsupials originated elsewhere before colonizing Australia from South America about 50 million years ago during the Eocene. In fact, the oldest fossil marsupials are *Kokopellia* from the mid-Cretaceous of Utah and *Deltatheridium* from the Late Cretaceous of Mongolia, and fossil evidence shows the presence of marsupials on all the Earth's continents in the geological past. Fortunately, it is possible to distinguish marsupials from placental

RIGHT Skull of the giant marsupial *Diprotodon optatum* from the Australian Pleistocene. Fossils of this animal were first found in caves in New South Wales and were brought to England where they were described in 1838 by Richard Owen.

mammals using features of their bones and teeth. For example, marsupials have three premolar and four molar teeth, whereas placental mammals typically have four premolars and three molars.

There are some remarkable examples of convergent evolution between marsupial and placental animals. Take the meat-eaters. The so-called Tasmanian 'wolf' *Thylacinus* is not a wolf but an Australian marsupial with a fossil record extending back to the Miocene, the last individual of which died in a zoo in 1936, while fossils of a marsupial 'lion' called *Thylacoleo* occur in the Miocene to Pleistocene rocks of Australia. Then there is a sabre-toothed marsupial, *Thylacosmilus*, which lived in South America during the Miocene and Pliocene epochs. *Thylacosmilus* sported even larger dagger-like canines than sabre-toothed placental mammals (p. 192). On the other hand, nothing quite like the Australian kangaroo has ever existed among placental mammals. The kangaroo *Procoptodon goliah* was huge, standing at least 2 m (6½ ft) tall and with the heaviest build of any known kangaroo. This species became extinct about 15,000 years ago and is possibly the large and aggressive kangaroo that attacked humans in the stories told by Aboriginal people.

Another extinct marsupial, *Diprotodon*, may have been recorded in Aboriginal legend as the Bunyip. Weighing an estimated 3,500 kg (3¾ tons), this Pleistocene mammal is the largest marsupial known to have lived. From a distance it might have resembled a brown bear. The huge chisel-like incisors of *Diprotodon* were employed to consume a broad range of plants, while bite marks on the fossil bones of *Diprotodon* show that it was preyed upon by the marsupial 'lion' *Thylacoleo*. However, the extinction of *Diprotodon* about 40,000 years ago was probably not brought about by *Thylacoleo* but by an extreme period of drought and fires as well as hunting by humans.

Placentals

Most mammals living today are placentals, mammals employing a placenta to transfer nutrients and waste matter between the blood of the mother and her foetus during gestation. While unequivocal fossils of placental mammals are lacking or extremely rare in the Mesozoic, molecular estimates and evolutionary modelling suggest that several groups of placentals were present during this era. Fossil evidence reveals spectacular diversifications of placentals belonging to numerous groups after the demise of the dinosaurs at the end of the Cretaceous.

AFROTHERIAN PLACENTAL MAMMALS

As the name suggests, afrotherians are a group of mammals with an African origin. They include today's elephants, sea cows and aardvarks, as well as hyraxes, elephant

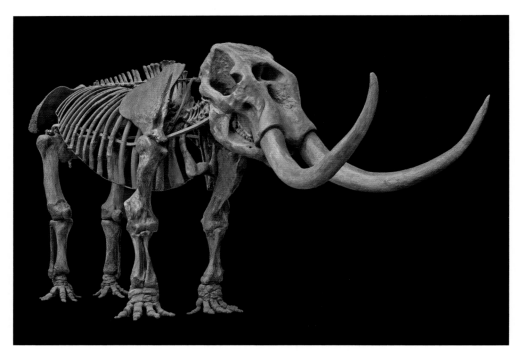

ABOVE Often confused with mammoths, this is a mounted skeleton of an American mastodon, *Mammut americanum*, excavated from Pleistocene deposits in Missouri, USA.

ABOVE Tooth from a Pleistocene woolly mammoth.

shrews, tenrecs and golden moles. The oldest known fossil elephants are Paleocene in age, whereas fossils of aardvarks date back to the Miocene and sea cows to the early Eocene.

Numerous species of large terrestrial mammals weighing over 45 kg (99 lb) have disappeared during the last 2.6 million years. The extinction of these animals, such as the woolly rhino (p. 194), giant deer (p. 197) and cave bears, has been attributed to both climate change and human impact, evidence for the latter including the coincidence between times of human colonization and extinctions. Mammoths epitomize this Pleistocene megafauna. The massive fossil teeth of these close relatives of elephants are commonly found in gravel pits, as are their tusks. In addition, many complete skeletons have been excavated, while mummified mammoths preserving hair, skin and internal organs are not uncommonly exhumed from the permafrost in Russia. The Pleistocene woolly mammoth *Mammuthus primigenius* was the last species of mammoth to become extinct, perishing perhaps as recently as 4,000 years ago. A further nine species of *Mammuthus* have been recognized. The oldest mammoth species is *Mammuthus subplanifrons* from South Africa, which is about 5 million years old, considerably pre-dating the Ice Ages with which these animals are usually associated. The largest species is the steppe mammoth *Mammuthus trogontherii*, which reached over 4 m (13 ft) in height. In contrast, the Sardinian mammoth *Mammuthus lamarmorai* stood just 1.4 m (4½ ft) tall, providing an example of island dwarfism. Several species of dwarf elephants once also inhabited islands in the Mediterranean. Skulls of these animals have a large central nasal opening resembling an eye socket, which may have led to their erroneous interpretation as skulls of the mythical one-eyed cyclops.

Sea cows, technically the order Sirenia, comprise today's dugongs and manatees. A mere four species of sirenians are extant. One dugong – Steller's seacow *Hydrodamalis gigas* – which fed on kelp in the Bering Sea, was hunted to extinction in the mid-eighteenth century just a few years after its scientific discovery. Sirenians have large tails, small front flippers, no hind flippers or external ears, and tiny eyes. The rib bones of sirenians are thick and dense, their weight helping these slow-swimming

BELOW *Arsinoitherium zitteli* was a horned mammal almost the size of a modern white rhino that inhabited North Africa during the Late Eocene and Early Oligocene.

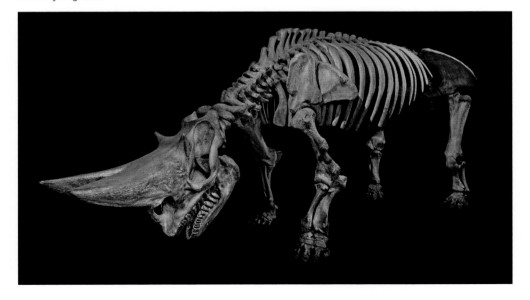

mammals to stay submerged. All modern sirenians are fully aquatic animals using up-and-down movements of the tail to swim, but some ancient sirenians had four well-developed legs, allowing them to walk on land. These include the Jamaican middle Eocene genus *Pezosiren*, which is believed to have employed its otter-like tail when swimming.

There are also some extinct afrotherian orders, among which is the Paleocene–Oligocene Embrithopoda. The best known embrithopod genus is *Arsinoitherium* named after Queen Arsinoe, the wife of King Ptolemy. *Arsinoitherium* was first found 80 km (50 miles) southwest of Cairo in the Fayum Depression, the palaeontological equivalent of the Valley of the Kings. Crocodile, turtle, bird and mammal fossils have been collected from more than 100 Eocene and Oligocene localities in the Fayum but none are more evocative than *Arsinoitherium*. Standing about 1.75 m (5¾ ft) tall, this stocky herbivore resembled a rhino and had a pair of large, bony horns fused at their base, with two small, knob-like horns behind them.

XENARTHRAN PLACENTAL MAMMALS

First appearing as fossils in rocks of Paleocene age, xenarthran mammals are native to the Americas and include just over 30 living species of sloths, armadillos and anteaters. All living sloths are arboreal mammals spending much of their time hanging from the undersides of branches. However, the extinct ground sloths had a different ecology, as is implicit in their name. The last ground sloths became extinct about 10,000 years ago after a 35-million-year-long history beginning in the late Eocene. These animals were often very large: the biggest was *Megatherium* from the Pleistocene of Patagonia, which weighed up to 4,000 kg (4 tons) and seems to have fed mostly on leaves using its long claws to pull down high branches.

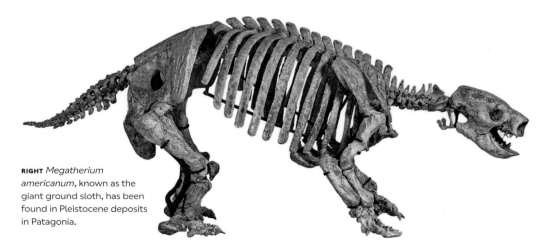

RIGHT *Megatherium americanum*, known as the giant ground sloth, has been found in Pleistocene deposits in Patagonia.

LEFT At 2.5 m (8 ft) long, this specimen of the armoured *Glyptodon clavipes* from La Plata in Argentina is of Pleistocene age.

An extinct genus of armadillo called *Glyptodon* is unmistakable because of the dome-shaped carapace made up of polygonal bony plates called osteoderms covering the body looking like a mammalian tortoise. *Glyptodon* was a large herbivore with a carapace measuring roughly up to 2.5 m (8 ft) in length. It was further protected by a shield of osteoderms over the head, and a tail encased in spiked osteoderms that could act as a defensive club against both would-be predators and competing individuals of the same species. Like the ground sloth *Megatherium* described above, hunting by human populations as they spread through South America probably played a major role in the extinction of *Glyptodon*.

LAURASIATHERIAN PLACENTAL MAMMALS

The third superorder of placental mammals is the Laurasiatheria. Again, the clue to the geographical origin of this group lies in the name, which refers to Laurasia, the northern landmass left when Gondwana drifted apart from Pangaea. Included in this diverse superorder are carnivores, many insectivores, bats, artiodactyl and perissodactyl ungulates and whales.

Carnivorous placental mammals are placed in three main orders – Carnivora, Hyaenodonta and Oxyaenodonta – the latter two orders formerly grouped together as Creodonta. Carnivora need little introduction. There are almost 300 living species of carnivorans, ranging from weasels to bears, wolves to lions, and walruses to seals. Their weights vary from 30 g to 1,000 kg (from 1 oz to 1.1 tons). Their fossil record extends back to the Paleocene and more than 350 fossil genera have been recognized.

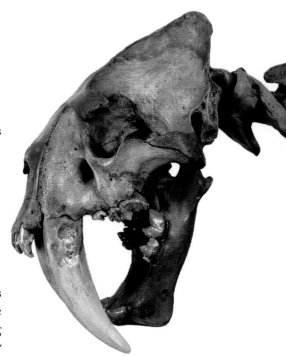

RIGHT The awesome canine teeth in the Pleistocene sabretooth *Smilodon fatalis* can be more than 17 cm (6¾ in) long. They are believed to have grown at a rate of about 7 mm (¼ in) each month.

Particularly striking among carnivorans are sabretooths with greatly enlarged upper canine teeth and short faces. No sabretooths are alive at the present day, although fossil evidence shows the repeated evolution of these dental dagger-wielding animals, not only among carnivorans but also marsupials (p. 187) and oxyaenodont creodonts (see below). The first carnivoran sabretooths are from the late Eocene where they are represented by *Dinictis*, a 1-metre-long (3¼ ft) animal belonging to the family Nimravidae. Sabretooth morphologies subsequently evolved in several genera belonging to the family Felidae, the true cats, which made their debut in the Oligocene. Undoubtedly the best-known sabretooth is *Smilodon*, star of several movies including *Ice Age* and *10,000 B.C.* At almost twice the length of *Dinictis*, *Smilodon fatalis* is represented by thousands of bones collected from La Brea in Los Angeles, California, USA. Individuals of this species inhabited the Los Angeles area over a period of more than 30,000 years during the Pleistocene epoch. Some were unwittingly trapped in

RIGHT An enormous skull of *Megistotherium osteothlastes* measuring 65 cm (25½ in) in length. This fossil was collected in 1966 by the leading fossil mammal expert Bob Savage (1927–1998) from Miocene deposits at Gebel Zelten in Libya.

LEFT Flattened, 'road-kill' fossil of the early bat *Palaeochiropteryx tupaiodon*. Measuring about 6 cm (2¼ in) across, this specimen comes from the Eocene oil shales of Messel near Darmstadt in Germany, a famous fossil Lagerstätte.

sticky tar exuding from the ground around watering holes. The tar later solidified into asphalt as the bones of *Smilodon* became fossilized.

Fossils of another carnivoran, the lion *Panthera leo*, reveal just how much the geographical range of animals can change through geological time. The oldest fossil lions come from the Pliocene of East Africa. By the Pleistocene, lions had spread into Europe, Asia and North America and possibly also Peru in South America. Remarkably, during the last interglacial period about 120,000 years ago, lions even inhabited what is now London, England – their fossil bones have been excavated from sediments deposited by the ancient River Thames at Trafalgar Square where, coincidentally, four Victorian sculptures of lions are arranged around Nelson's Column. Lions today are confined to sub-Saharan Africa, apart from a single population in India.

'Creodonts' were also carnivorous placentals appearing at about the same time as carnivorans and becomng extinct during the Miocene. They evolved to become larger and more diverse than carnivorans in the early Eocene but dwindled thereafter. Often more squat in posture and shorter limbed than other carnivorans, creodonts differ in several characters of the bones and teeth. The hyaenodontan creodont *Megistotherium* from the Miocene of Saharan Africa had a skull that was 65 cm (25½ in) long, twice the size of a tiger's skull. It may have been the largest carnivorous land mammal ever to have lived.

The oldest fossils of bats (Chiroptera) have been found in rocks of Eocene age when the mostly insectivorous, echolocating microbats made their debut. The fruit-eating megabats ('flying foxes') have a poorer fossil record but are believed to have originated in the Oligocene. More than 70 taxa of Eocene bats have been recognized, distributed across five continents (North America, Europe, Africa, Asia and Australasia). Evidence points to

RIGHT The elongate skull of the woolly rhinoceros *Coelodonta antiquitatis* can reach 80 cm (31½ in) long. This animal lived during the Pleistocene Ice Ages in Europe and Asia, feeding on grassland sedges growing on the cold steppe.

bats undergoing an explosive evolutionary radiation at this time, roughly 50 million years ago. Contrary to expectations, echolocation seems not to have triggered this radiation as fossil ear bones suggest the ability to echolocate was acquired somewhat later and well after flight had evolved. Bats are common fossils in some Pliocene to Recent cave deposits where their bones are buried in thick deposits of their own phosphate-rich droppings.

Ungulates are hoofed mammals. There are two main living ungulate orders: Perissodactyla, which contain horses and rhinos; and Artiodactyla, to which cattle, pigs, giraffes, camels, deer and hippopotamuses belong. Through the Cenozoic the relative abundance of these two ungulate orders changed, from an early dominance of perissodactyls to the overwhelming dominance of artiodactyls seen today.

Only six genera of perissodactyls live today, this once diverse group having declined since the Eocene. Initially lacking horns, rhinos underwent an evolutionary radiation in the Eocene but have dwindled in diversity towards the present day. Among more recent rhinos are *Elasmotherium*, a Pleistocene giant from the Siberian steppes that sported a single horn up to 2 m (6½ ft) in length, and the Ice Age woolly rhino *Coelodonta antiquitatis*. The woolly rhino ranged widely across Europe and northern Asia, featuring in Late Palaeolithic cave paintings. Some believe this animal to have been the inspiration for the mythical dragon – its geographical distribution certainly corresponds with that of dragons, while the elongate skull resembles those of some depictions of dragons.

The evolution of horses (family Equidae) is a favourite textbook example of evolution in the fossil record. While the old idea of a single lineage leading to modern horses is no longer tenable, there are clear trends through time in the stature, skeletal structure and cheek teeth of horses that correspond to changes in habitat. Horses have

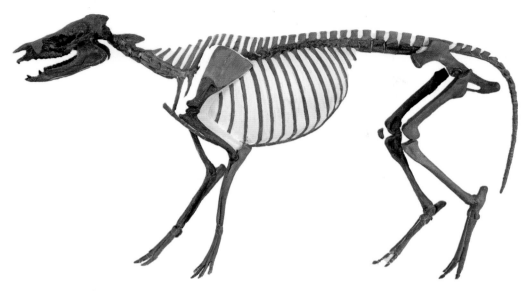

ABOVE Skeleton of the perissodactyl *Hyracotherium leporinum* measuring 80 cm (31½ in) in length. Fossils of this horse-like animal were first found in the Early Eocene London Clay of Herne Bay in Kent, England.

increased in size, while their feet have been formed by ever fewer toes and their teeth have become more elongate as they transitioned from browsers to grazers. The early Eocene genus *Hyracotherium* was first found in the London Clay of Kent, England, in the early nineteenth century. It is now recognized as a palaeothere, a close relative of horses. The oldest true horse is the earliest Eocene *Sifrhippus* from North America. It was not much bigger than a domestic cat, had four toes on the front feet and three on the back, and was probably a woodland browser using its low, enamel-covered teeth to feed on fruits, seeds and leaves. *Mesohippus* from the late Eocene to the early Oligocene was about the size of a sheep, had three toes on all feet, and is believed to have fed on mixed vegetation in more open environments. Toe number was reduced to one in the late Miocene *Dinohippus*, which stood almost as tall as a present-day horse. The teeth of *Dinohippus* are high-crowned and have a covering of cement to help resist abrasion from grit in the soil and silica in the grasses during grazing. The genus *Equus*, which includes not only modern horses but also asses and zebras, first appeared in the Pliocene. *Equus ferus*, the species to which domesticated horses belong, may have appeared a little over a million years ago. Interestingly, *Equus ferus* originated in North

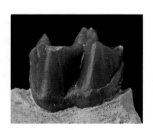

ABOVE Molar tooth of the horse *Mesohippus bairdii* from the Oligocene of Munson Ranch near Harrison in Nebraska, USA. The tooth measures less than 1.5 cm (½ in) across.

RIGHT Skull of the oreodont *Merycoidodon culbersoni* from the Miocene of White River in Dakota, USA. Measuring 26 cm (10¼ in), this example was presented to the Natural History Museum, London by American palaeontologist Joseph Leidy (1823–1891).

America, dispersed into South America, Asia, Europe and North Africa, then became extinct in North America about 10,000 years ago and was subsequently re-introduced into this continent by humans.

Most domestic livestock are artiodactyls. These ungulates have simpler teeth than perissodactyls but more complex stomachs where processing of food occurs slowly. The earliest artiodactyls were small animals with long tails. They were of early Eocene age, other groups appearing later in the Eocene and during the Oligocene or Miocene. Among extinct families are the North American oreodonts, fossils of which have been collected in huge numbers from Oligocene and Miocene rocks in the Badlands

BELOW The huge antlers of this specimen of *Megaloceros giganteus* measure 3.2 m (10½ ft) from tip to tip. Collected from Pleistocene deposits in the Bog of Ballybeta, Ireland, this giant extinct deer is often erroneously referred to as the 'Irish Elk'.

ABOVE Pleistocene premolar tooth of the extant hippo *Hippopotamus amphibius*. Measuring 3.7 cm (1½ in) across, this fossil tooth was collected from East Sussex in the south of England.

ABOVE RIGHT Fossil ear bone of a whale found in the Plio-Pleistocene Red Crag Formation of Felixstowe in Suffolk, England. The specimen measures about 12 cm (4¾ in) in size.

of South Dakota and Nebraska. The oreodont *Merycoidodon* was the size of an average pig or sheep, had a full complement of cheek teeth, and apparently roamed in herds browsing on vegetation.

Deer (cervids) are characterized by the presence in males of deciduous antlers. In the earliest cervids from the Miocene epoch, the antlers tend to be simple compared to those of more recent cervids which are generally larger animals. One of the grandest is the giant deer *Megaloceros giganteus*, which appeared during the Pleistocene and became extinct less than 10,000 years ago. This majestic animal stood 2 m (6½ ft) high at the shoulders and had outrageously large antlers, sometimes measuring more than 3.6 m (11¾ ft) from antler tip to antler tip! Many antlers of *Megaloceros* were recovered from Irish bogs during the nineteenth century and purchased by aristocrats for display alongside the puny antlers of the contemporary deer they had shot.

Hippos are represented today by only two species, both living in Africa, but the family Hippopotamidae was more diverse and geographically widespread in the geological past. First recorded in late Miocene rocks of Kenya, hippos subsequently spread beyond Africa into Eurasia. Fossil hippos belonging to the extant species *Hippopotamus amphibius* are even known from Pleistocene deposits in Britain.

There is no mistaking a cetacean, the mammalian infraorder that includes whales, dolphins and porpoises and is particularly notable for the blue whale, which – at almost 30 m (99 ft) in length – is the largest animal known to ever have existed. Often considered as the most intelligent of non-primate mammals, over 80 cetacean species live today. They are divided between two groups, odontocetes (toothed whales including dolphins) and mysticetes (baleen whales). Fossil cetaceans are moderately

RIGHT Skull of the Late Oligocene whale *Janjucetus hunderi* from Jan Juc in Victoria, Australia. This primitive baleen whale, known from only a single specimen, is estimated to have had a body length of about 3.5 m (11½ ft).

BELOW Believed to be a semi-aquatic forerunner of whales, this fossil of *Pakicetus attocki* was found in the Eocene rocks of Pakistan. Individuals of *Pakicetus* reached 1–2 m (3¼–6½ ft) in length.

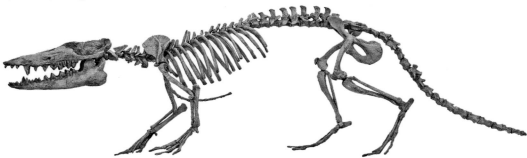

rare but tend to be conspicuous when found. External parts of whale ear bones called tympanic bullae are not uncommon, but the scarcity of more complete specimens can be explained in part by the fact that whale carcasses often sink to the seabed in the deep ocean rather than in the shallow water environments that are better represented in the sedimentary rocks exposed on land. An exception is a Miocene locality called Cerro Ballena in Chile, where more than 30 whale skeletons have been found at four levels. These seem to be the result of mass mortalities after toxic algal blooms poisoned the animals, whose carcasses then drifted ashore.

Strictly speaking, whales are artiodactyl ungulates, closely related to the hippo family. The evolution of whales from a terrestrial to a fully aquatic lifestyle is well seen in fossils of Eocene age. The closest known fossil relative to the cetaceans is an even-toed ungulate (artiodactyl) called *Indohyus*. Bones of this small deer-like animal have been found in earliest Eocene rocks in India. The middle ear shares characteristics with those found in whales, while the thick cortical layer of the limb bones recalls some modern wading mammals, evidence that *Indohyus* spent a significant amount of time in the water. More whale-like than *Indohyus* is the middle Eocene *Pakicetus*, which may have been an amphibious, semiaquatic animal feeding in the water.

The earliest true cetaceans, the so-called archaeocetes, include a middle to late Eocene group called basilosaurids that were fully aquatic marine animals more similar in appearance to modern cetaceans. *Basilosaurus* itself is the state fossil of both Alabama and Mississippi. It reached 18 m (59 ft) in length and fed on fish, judging from fossilized stomach contents. Unlike modern whales, however, *Basilosaurus* had small hind legs and the skull was much shorter in proportion to the body. Archaeaocetes became extinct at the end of the Eocene, and subsequent cetaceans belong to either the odontocetes (toothed whales such as the killer whale and dolphins) or mysticetes (baleen whales such as the blue whale). Odontocetes today employ underwater echolocation and there is fossil evidence that this ability had evolved by the Oligocene. Dolphins appear a little later in the Miocene. As for mysticetes, the oldest fossil example, *Mystacodon*, was found in late Eocene rocks in Peru. As with the younger genus *Janjucetus* from the Oligocene, *Mystacodon* has conventional teeth and not the plates of keratin called baleen used by modern mysticete whales for filter-feeding. It appears that baleen evolved later and is now ubiquitous among mysticete whales.

EUARCHONTOGLIREAN PLACENTAL MAMMALS

The superorder Eurachontoglires consists mainly of rodents, rabbits (lagomorphs) and primates. With Laurasian geographical origins, this group probably originated in the Late Cretaceous based on molecular estimates. However, no definite fossil record is known until the Paleocene when the earliest rodent and primate fossils appeared, with lagomorphs perhaps dating back in the fossil record to the Eocene.

As mentioned previously, rodents are the most diverse group of living mammals. They are characterized by possessing only one pair of incisor teeth in each jaw, these

RIGHT Cast of the skull of the Oligocene primate *Aegyptopithecus zeuxis*. Measuring about 10 cm (4 in), this fossil was collected from Oligocene rocks exposed in the Fayum Depression of Egypt.

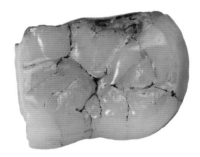

ABOVE Broken molar tooth of *Gigantopithecus blacki* from a cave in Lincheng, China. This large Pleistocene ape has been associated with the yeti legend even though evidence points to its extinction more than 200,000 years ago.

teeth continuously growing and being worn down with use. Modern rodents exhibit a great range in size, and rodent variation is even greater when fossil species are taken into account. The largest living rodent is the South American capybara, *Hydrochoerus*, tipping the scales at 50 kg (110 lb), whereas the Brazilian Miocene genus *Telicomys* may have been 10 times this weight and attained the height of a small rhino.

Rabbits belong to the order Lagomorpha, together with hares and pikas. The majority are herbivores adapted to fast running and leaping to evade predators. About 110 species of lagomorphs live today, with more than 230 fossil species, mostly represented by teeth or fragments of jaws. More complete skeletons have been found of the North American Eocene–Miocene genus *Palaeolagus*. This genus closely resembled a small modern rabbit, and indeed lagomorphs seem not to have changed greatly in morphology since their inception, except in their teeth, which now grow continuously.

The oldest possible primate fossil comes from the late Paleocene of Morocco, and within the first few million years of the Eocene primates began an evolutionary radiation in Europe, Asia and North America. An outstandingly preserved fossil of an Eocene primate, nicknamed 'Ida', was discovered in 1983 at Messel in Germany. The articulated skeleton of *Darwinius massillae* is about 58 cm (22¾ in) long, and represents a juvenile female. Probably having an arboreal lifestyle, the stomach contents of Ida show that it fed on fruit and leaves. The Oligocene primate *Aegyptopithecus* was first discovered in the Fayum Depression of Egypt (p. 190). It belongs to the group from which both apes and Old World monkeys evolved. It was an estimated 70 cm (27½ in) in length and studies of wear patterns on the teeth suggest that *Aegyptopithecus* subsisted on a diet mostly of fruit.

More than 50 genera of apes have been described from the Miocene. Indeed, the Miocene has been dubbed the true 'Planet of the Apes'. Yet it has been estimated that only about 1% of primate species that lived during this epoch are actually known as fossils (this estimate assumes a duration of about 1 million years for each species and a similar diversity at each moment in time to that of today). All early Miocene apes lived in Africa but in the middle Miocene they expanded their range into Europe and Asia. Great apes did not survive in Europe beyond the Miocene.

A particularly large fossil primate was *Gigantopithecus*, which occurs in the Pleistocene of China. This ape was made known to science by Gustav von Koenigswald (1902–1982), who purchased a tooth in 1935 from a pharmacy in Hong Kong. The molar teeth of *Gigantopithecus* are the largest known in any ape. However, the lack of complete skeletal remains hinders an accurate estimate of the size of this animal. Estimates suggest it may have weighed as much as 300 kg (660 lb), almost half as much again as a modern gorilla. Analysis of proteins from the tooth enamel of *Gigantopithecus* show its closest living relative to be the orang-utan.

Hominins

Modern humans and other hominids are also primates. According to molecular estimates, modern humans (*Homo sapiens*) share with our closest living relatives (chimps and bonobos) a common ancestor that lived between 6 and 7 million years ago in the late Miocene. This is when the branch of the evolutionary tree containing modern humans and various extinct hominins spilt off from the branch containing chimps and bonobos. It was once believed that the hominin branch consisted of a single lineage. According to this view only one species of hominin existed at any point in time and each newly discovered fossil hominin could be justifiably hailed as a 'missing link' on the way to *Homo sapiens*. With the discovery of increasingly more fossil hominins, including species that clearly lived at the same time (if not necessarily in the same place), the single lineage hypothesis of human evolution has been abandoned. Instead, hominins constitute a branching tree, as yet of uncertain topology, containing extinct side branches as well as the solitary surviving branch that terminates in *Homo sapiens*.

While the casual fossil collector is very unlikely to stumble across a fossil of an extinct hominin, targeted excavations around the world have yielded a huge assortment of hominin fossils, the oldest coming from East and South Africa. Most of these fossils are incomplete – many are fractured bones, broken skulls or just isolated teeth. Some occur in cave deposits, the unfortunate individuals either perishing after falling through holes in the cave roof, or being dragged into the cave by predators or scavengers. Other hominin fossils are found in sediments deposited in the flood plains of rivers and lakes, as at the famous Oldupai Gorge (formerly Olduvai Gorge) in Tanzania, where the celebrated Leakey family and their collaborators made some remarkable finds during the second half of the twentieth century. In addition to bones and teeth, cultural artefacts, notably stone tools, add an important dimension to our understanding of human evolution. Then there are footprints, a kind of trace fossil (see Chapter 13), like those discovered in 2013 at

RIGHT Skull of a young individual of *Australopithecus sediba*. This small hominin was discovered in the Malapa cave system in South Africa and is of early Pleistocene age.

Happisburgh in Norfolk, England, which were made by humans crossing a muddy riverbank almost 1 million years ago.

Several fossils have been proposed as the oldest known hominins. Among these the most complete skeleton belongs to the 4.6–4.3-million-year-old *Ardipithecus ramidus* from Ethiopia. With a height of 1.3 m (4¼ ft), this hominin had a chimp-sized brain but probably walked bipedally, while retaining adaptations for climbing trees such as an opposable big toe. Seven species have been assigned to a slightly younger genus called *Australopithecus*, which lived from about 4.2 to 2 million years ago. The size of the cranium indicates that *Australopithecus* had a larger brain than *Ardipithecus*, and the evidence of a 3.7-million-year-old footprint site at Laetoli in Tanzania (see p. 245), almost certainly made by *Australopithecus*, suggests a bipedal gait very similar to that of modern humans. *Australopithecus* has larger molars than chimps but not nearly as massive as those of another hominin, *Paranthropus*, which lived from about 2.6 to 1.3 million years ago. Indeed, one fossil of *Paranthropus* from Oldupai Gorge was nicknamed 'Nutcracker Man' alluding to the huge cheek teeth with a thick layer of enamel. At the time it was discovered, this was interpreted as an adaptation for dealing with hard food items, but more recent isotopic evidence points to a diet of grasses or sedges that necessitated the processing of large quantities of this low-quality food.

This brings us to the genus *Homo*, the first and only hominin known to have spread beyond the continent of Africa. At least a dozen different species of *Homo* are now recognized. Many scientists are happy to consider all of these as humans of a sort. By this reckoning, the oldest known fossil of a human dates back approximately 2 million years, whereas the oldest fossil of *Homo sapiens* is about a quarter of a million years old. Brain size in *Homo* is generally larger than in *Australopithecus*, perhaps explaining the ability of *Homo* to manufacture stone tools. Some of the variation encompassed within the genus *Homo* is clear when comparing *Homo rhodesiensis* with *Homo floresiensis*. The first of these species is represented by a superbly preserved skull discovered at Broken Hill in Zambia and dated to about 300,000 years. It belongs to a geographically widespread group that also includes *Homo heidelbergensis*, some examples of which may be more than half a million years old. These species were broad-shouldered hominins with heavy brow ridges. They grew to a similar height as modern humans and their brains were about the same size. In contrast, *Homo floresiensis* measured just over a metre in height, leading to its nickname the 'hobbit', and had a brain roughly the same size as a chimp's. First discovered as recently as 2003 in a cave on the Indonesian island of Flores, *Homo floresiensis* seems to be part of a group of diminutive humans that colonized this island perhaps as long as 700,000 years ago, persisting there for more than half a million years.

The young age and fine preservation of some fossil species of *Homo* has allowed their DNA to be extracted, sequenced and compared with that of modern *Homo sapiens*. This has provided evidence of occasional hybridization between contemporaneous species of *Homo*. For example, modern human genomes can contain a small percentage of genes inherited from the stocky, cold-adapted Neanderthals (*Homo neanderthalensis*) that became extinct about 40,000 years ago. Despite this genetic mixing, it is relatively easy to distinguish *Homo sapiens* anatomically from extinct species of *Homo* by the flat forehead lacking brow ridges and the protruding chin.

BELOW Four views of the skull of *Homo rhodesiensis*. Estimated to be 300,000 years old, this Middle Pleistocene fossil was collected in 1921 at the Broken Hill mine in Zambia.

11 Fossil plants: the greens join the party

It has been estimated that plants account for 80% of the biomass on Earth today. Plants are vital to the survival of all life on our planet – and they are also crucial to our understanding of how our planet's life evolved. Through the process of photosynthesis, plants use sunlight, carbon dioxide and water to manufacture complex organic molecules used to construct leaves, shoots and roots – the food of herbivorous animals, which in turn are eaten by carnivores. At the same time, photosynthesis generates oxygen for animals to breathe and removes from the earth's atmosphere the greenhouse gas carbon dioxide, whose increase in recent years has precipitated the global climate crisis.

Plants are predominantly multicellular organisms with cells possessing nuclei enclosed by membranes (i.e. eukaryotes) as well as chloroplasts, the organelles in which photosynthesis occurs. As mentioned earlier (see Chapter 2), plants were not the first organisms to employ photosynthesis: prokaryote cyanobacteria acquired this ability billions of years before plants had evolved. In fact, the chloroplasts of plants were sequestered from cyanobacteria, becoming incorporated into the host cells as symbionts. Subsequent milestones in plant history include the colonization of terrestrial environments ('the greening of the land'), the evolution of trees, and the origin of flowers. Fossil plants tell us how and when these events occurred.

They are also a tremendous source of information about ancient environments: the study of growth rings in trees sheds light on seasonality; leaf shapes in flowering plants reflect climatic conditions. For the latter, the degree to which their margins are toothed is proportional to mean annual temperature, and leaf size to mean annual rainfall. Moreover, the density of stomata (pores controlling gas exchange) correlates with carbon dioxide levels. We know a lot about the environments favoured by present-day plants and it is likely that their close fossil relatives flourished in similar conditions. This is another method –uniformitarianism – that uses fossil plants to interpret palaeoenvironments.

OPPOSITE A beautiful fossil of *Neuropteris* from the Carboniferous. This plant is a pteridosperm, otherwise known as a 'seed fern'.

How are plants fossilized?

Unlike animals, land plants do not have biomineralized skeletons, and yet their fossils are exceedingly abundant, which is largely a testimony to their rigid cell walls. Plant fossils can be preserved in several different ways. The commonest modes of preservation are as compressions and impressions – compressions retain some of the plant tissues in a flattened form, whereas impressions are represented by voids or imprints in the sediment with no tissue preservation.

Plant tissues are formed using a range of organic compounds. These include the carbohydrate cellulose and lignin in the cell walls, and lipids in the waxy external cuticle, all of which are relatively resistant to decay. Nevertheless, even these resilient compounds degrade through time, gradually transforming to a carbon-rich insoluble material known as kerogen. This process of carbonization accounts for the widespread preservation of fossil plants as black films. Other processes play their part in fossil plant preservation. A thin layer of hydrous iron oxide minerals may cover the surfaces of compression and impression plant fossils, especially leaves, that were deposited in aqueous environments. These rusty coatings are attributed to biofilms of bacteria developing on the waxy plant cuticle and promoting the precipitation of a mineral called ferrihydrite. Exquisite tiny flowers and the cellular structures of other plant organs can be preserved as fusain, fossil charcoal resulting from the burning of plants in natural wildfires, or ignited by lava flows or hot volcanic ash falls. Plant materials preserved in this way are exceedingly delicate and require special conservation.

An entirely different mode of preservation – petrifaction or permineralization – is associated particularly with fossil wood. Here, inorganic minerals grow inside the cells and other spaces within the plant, often preserving the finest anatomical details. Petrifaction occurs at, or soon after, burial and usually involves precipitation of silica (chalcedony), but sometimes calcium carbonate minerals or pyrite. Replacement of the cell walls by the same minerals often accompanies permineralization. Not all fossil wood is permineralized: 'mummified' wood is unmineralized and has undergone minimal degradation, usually with the lignin in the cell walls surviving. Some remarkable examples of 'mummified' wood have been described in deposits as old as 60 million years. For example, at Ipolytarnóc Fossil Forest in Hungary, Miocene stumps of the tree *Glyptostroboxylon* were mummified when a rising lake drowned a forest, rapidly burying the trees in sediment saturated in water, impeding normal processes of decay.

Early mineral growth in concretions can preserve the external form of fossil plants, as in the famous Carboniferous Mazon Creek ironstone concretions of Illinois, USA. More than 400 different species of fossil plants have been described encased in these concretions, which were formed by the growth of inorganic siderite (iron carbonate), often centred around decaying plants and animals, in the sediments deposited by a large system of deltas.

Plants are seldom preserved in their entirety. Instead, different parts typically become dissociated from one another after the plant has died – leaves and flowers detach from branches or fall apart, stems break away from roots etc. These individual organs are typically given different taxonomic names. For example, a common tree inhabiting Late Carboniferous Coal Measure forests is the source of fossils known as *Lepidodendron* (tree trunks), *Stigmaria* (roots), *Lepidophylloides* (leaves) and *Lepidostrobus* (cones). The discovery of occasional specimens showing organic connection between these 'fossil-genera' makes it clear that they are all parts of the same plant. The rules governing the naming of plants allow for different names to be retained for fossils representing different parts of the same plant, meaning that all four of the generic names quoted above, for example, are still in use.

The first land plants

The evolution of the terrestrial land plants that today carpet our world was one of the greatest events in geological history. It had profound consequences for all living organisms and the kinds of environments available for them to inhabit. Molecular evidence points to aquatic green algae related to the filamentous pondweed *Spirogyra* as the ancestors of land plants, and it is likely that simple green algae preceded complex plants into terrestrial environments perhaps 600 million years ago. Modern land plants – classified as Embryophyta – fall into two major groups: bryophytes and vascular plants. Vascular plants differ from the more basic bryophytes in having elaborate conducting tissues to transport water and minerals (xylem) and the products of photosynthesis (phloem). Living bryophytes, numbering about 22,000 living species, are mostly mosses and liverworts. Neither of these two groups is well represented in the fossil record because of their fragile cell walls. However, fossils thought likely to be liverworts have been described back to the Devonian, and mosses are known in rocks as old as Carboniferous in age, but fossils before the Mesozoic are rare. After this, pieces of moss encapsulated in amber become increasingly common.

The oldest widely accepted fossils of 'adult' land plants come from the mid-Silurian, although tiny spores possibly from land plants are known from several Cambrian and Ordovician sites, suggesting an earlier history not evident from macrofossils alone. The standard bearer among early plants is *Cooksonia*, which has been found globally in rocks as old as 432 million years. This tiny plant grew to just a few centimetres in height and consisted of simple stalks dividing a few times, each stalk capped by a trumpet-shaped sporangium housing the spores. It had no leaves. A waxy, waterproof cuticle prevented *Cooksonia* from drying out, and stomata are present to allow gas exchange. Importantly, one of the younger species of *Cooksonia* shows some indication of vascular tissues in the stalks, providing a link between bryophytes and fully vascular plants.

Mentioned already in the context of the first insects (p. 58), the Rhynie Chert is a remarkable Early Devonian deposit from Aberdeenshire, Scotland that was discovered in 1912 by physician William Mackie. As the earliest in situ terrestrial ecosystem known from anywhere in the world, the plants of the Rhynie Chert provide precious insights into early vascular plants existing some 407 million years ago. The Rhynie fossils have an unusual origin. Animals, plants and fungi living close to a hot spring were periodically inundated with hot water charged with silica, a process replicated at the present day in geothermal areas such as Rotorua in New Zealand. The entombed plants were permineralized in silica, preserving cellular structures that are revealed in their full glory when thin sections are made of Rhynie Chert. All seven species of Rhynie plants have simple branching stalks lacking true leaves and roots, although they did possess

ABOVE This simple land plant *Cooksonia pertoni* has spore capsules at the ends of the bifurcating branches. Measuring about 1 cm (½ in) in length, this fossil was found in the Late Silurian rocks of Herefordshire, England.

ABOVE RIGHT A thin section of Early Devonian Rhynie Chert intersects the permineralized fossil of the plant *Asteroxylon mackiei*, with leaf-like structures visible on the left-hand side of the stem. The field of view in this image is about 8 mm (¼ in) across.

subterranean rhizomes and filament-like rhizoids. Crucially, the Rhynie plants have vascular tissues: water-conducting cells forming a narrow xylem. However, the presence of a phloem to transport the products of photosynthesis is less well established.

Lycopods and the coal forests of the Carboniferous

The Silurian and Devonian plants discussed so far are unlikely to feature in many fossil collections, in contrast to plants from the succeeding Carboniferous period, especially the Late Carboniferous, or Pennsylvanian as it is called in North America. The Late Carboniferous was a time when vast forests swathed equatorial regions in many parts of the world. These contained plants of all shapes and sizes. A lot is known about this luxuriant vegetation because of commercial mining of plant-bearing deposits for coal. No geological period has witnessed as much coal formation as the Carboniferous when tropical forests stretched across the continent of Pangaea from Kansas to Kazakhstan. Although these forests have been compared to modern tropical rain forests such as the Amazon, they were different in many ways, not least in the kinds of plants present. The Carboniferous flora was dominated by lycopods ('clubmosses'), ferns, pteridosperms ('seed ferns'), equisetales ('horsetails') and cordaitales (relatives of conifers). Apart from ferns, these groups are either extinct (pteridosperms and cordaitales) or are now pale shadows of what they were in the Carboniferous (lycopods).

Each of these plant groups contributed to the formation of Carboniferous coal in wet, swampy environments – peat mires – where low levels of oxygen retarded decomposition of organic remains. Burial beneath new layers compacted the accumulating peat, with loss of water, methane and carbon dioxide leading to a corresponding increase in elemental carbon content. Peat transitioned to brown coal, and brown coal to black bituminous coal and anthracite, depending on the temperatures and pressures experienced. Sequestration of so much carbon during the Late Carboniferous led to a rise in atmospheric oxygen levels and with it the likelihood of wildfires. Some coals contain the residues of these fires in the form of fossilized charcoal known as fusain. Concretions known as coal balls are occasionally found in coal-bearing rocks. Plants contained in coal balls are permineralized with calcium carbonate, preserving cellular details that are not usually seen and contributing greatly to our knowledge of the structure and biology of these residents of the first tropical forests.

Lepidodendron is one of the most familiar plants of the Carboniferous coal forests. Compression and impression fossils of the bark of this lycopod are common on spoil tips from coal mines. Covered by distinctive diamond-shaped scars left after detachment of the leaves, these fossils can look at first like the scaly skin of reptiles. *Lepidodendron* grew as a tree, attaining a height of 40 m (131 ft) and

RIGHT The black colour of this fossil of the lycopod *Lepidodendron subdichotum* is typical for a specimen from the Late Carboniferous Coal Measures. Measuring 6.5 cm (2½ in) at the base, each scale on the branch marks the attachment point of a long leaf.

FAR RIGHT *Stigmaria ficoides* represents the fossilized roots of large lycopod trees such as *Lepidodendron*. This 15 cm (6 in) long specimen from the Carboniferous shows the dimples where narrow rootlets were once attached.

exceeding 1 m (3¼ ft) in diameter at the base. Initially comprising a single stem, *Lepidodendron* trees of a certain height developed a profusely branched canopy, many of the branches terminating in a cone known as *Lepidostrobus* containing spores for reproduction. *Lepidodendron* trees contained a surprisingly small amount of true wood considering their large size. However, their bark was very thick and 'woody', providing most of the support the trees needed to grow tall. Up to 45% of the plant tissue in Carboniferous coals may consist of bark from *Lepidodendron*, *Sigillaria* and their close relatives.

The stumps of 11 *Lepidodendron* trees with attached root systems can be viewed at the geological heritage site of Fossil Grove in Victoria Park, Glasgow, Scotland. Remarkably, the tree stumps are upright and in their original life positions. Their preservation entailed several stages. First, the swampy environment in which the trees grew was flooded, killing the trees and depositing sediment around them. Next, the rotting trunks toppled over, and the stumps and roots were hollowed out through decay while the tough outer bark remained intact. More sandy sediment was washed into the hollow stumps and roots, hardening with time into sandstone. Thus, the tree stumps are preserved as sandstone casts wrapped by the carbonaceous remnants of the bark. As for the roots, these too are sandstone casts of a root type commonly encountered in the Carboniferous and called *Stigmaria*. Lycopod trees were anchored by extensive root systems. Spirally arranged circular scars on specimens of *Stigmaria* represent the positions of rootlets that were highly branched and provisioned with hairs as in many modern plants.

Horsetails were another group of tree-forming Carboniferous plants. At the present-day horsetails are represented by just a single herbaceous genus (*Equisetum*), which can be found growing as a weed at the sides of railway lines and roads in Britain. *Equisetum* is characterized by having needle-like leaves arranged in whorls around a hollow stem. However, during the Carboniferous horsetails were more diverse and varied. *Calamites* grew as trees reaching a height of 10 m (33 ft) in the Late Carboniferous, although these horsetail trees were less common than lycopod trees in peat mires and instead tended to be found along the edges of lakes and streams. Fossil branches of *Calamites* are striated along their lengths and have annular leaf whorls like those of living *Equisetum* plants. An entire *Calamites* tree would have vaguely resembled a Christmas tree.

Ferns inhabiting the Carboniferous peat mires were broadly similar to modern ferns, some growing at ground level but others forming small trees. Most are compression or impression fossils of fronds, with an axis giving rise on both sides to lateral branches bearing numerous small leaves (pinnules), a morphology familiar among ferns living today. Another group, the pteridosperms, had foliage that could easily be mistaken for a fern. However, rather than possessing spores on the undersides of the pinnules, these extinct plants produced seeds for reproduction, hence the common name 'seed fern'. Some pteridosperms also grew as trees, but these contained woody tissues in contrast to tree ferns, which have trunks formed from the fibrous bases of the leaves. Pteridosperms were the dominant plants growing on the flood plains of some Late Carboniferous rivers. *Neuropteris* is a widespread fossil of a pteridosperm found in Carboniferous rocks. It is not uncommon to find *Neuropteris* pinnules with excisions along the pinnule margins believed to have been made by predatory insects.

ABOVE Approximately 40 cm (15¾ in) in length, this fossil is the horsetail *Calamites* from the Late Carboniferous. The curved base of the specimen is probably where the vertical stem transitioned to a horizontal root.

Pteridosperms are classified within the gymnosperms (plants with naked seeds, cf. angiosperms with seeds enclosed within fruits) along with Cordaitales, another extinct order containing tree-forming species. Resembling conifers in their dense wood, cones and seeds, these plants had long, strap-shaped leaves unlike conifer needles. Trees belonging to the genus *Cordaites* grew up to 30 m (99 ft) tall and had a single unbranched trunk. While some *Cordaites* trees lived in Carboniferous peat mires alongside lycopods and tree ferns, others seem to have been elements of upland floras.

ABOVE Mass of leaves from two species of *Glossopteris* trees cover a piece of rock 60 cm (23½ in) long from the Late Permian of Nagpur, India.

RIGHT Leaf of the Late Carboniferous gymnosperm tree *Cordaites borassifolius*. This fossil measuring about 40 cm (15¾ in) was collected at Cannelton in Pennsylvania, USA.

Commercially mined coals in what is now the northern hemisphere were formed mainly during the Carboniferous period, but coals mined in India, southern Africa, South America and Australia are generally younger in age, dating from the Permian period. The dominant Permian plants in the southern continent of Gondwana were *Glossopteris* and its close pteridosperm relatives. *Glossopteris* trees could grow up to 30 m (99 ft) tall and were deciduous, shedding huge numbers of long, tongue-shaped leaves, which are commonly found as fossils. These woody trees inhabited forests in lowland swampy mires in a similar way to the coal-associated Carboniferous plants described above. However, these peat mires were formed at higher, temperate latitudes, not in the tropics. A famous find of *Glossopteris* leaves during the tragic final expedition to Antarctica of explorer Robert Falcon Scott shows that these trees could even grow at ancient latitudes greater than 80°S.

It is worth remarking that the lycopod and the other vascular plant trees mentioned above were not the first trees on the surface of our planet. A curious Devonian fossil called *Prototaxites* was able to construct trees which reached almost 9 m (29½ ft) in height and a metre in diameter. Formed from mats of filaments, the identity of *Prototaxites* is contentious. However, the prevailing view is that – astonishingly – it was the fruiting body of a kind of fungus!

Gymnosperms and other Mesozoic plants

The Mesozoic was a time when herbivorous dinosaurs consumed vast quantities of vegetation to meet their gargantuan dietary requirements. Gymnosperms and other non-flowering plants were all that was on offer for most of this era. The flowering angiosperms – both broad-leaved and grasses, which are so important in the world today – were scarce until the middle of the Cretaceous period some 100 million years ago. Common Mesozoic plant fossils include conifers, cycads, bennettitaleans, seed ferns (pteridosperms), ferns, horsetails and *Ginkgo*.

Most fossil wood comes from conifer trees. Conifers are a division of gymnosperms called Pinophyta. There are many sites around the world where fossil conifer wood has been found, but few rival the Petrified Forest National Park in Arizona, USA. Numerous permineralized trees occur here in the Late Triassic Chinle Formation. So numerous are the trees that they were used by indigenous people in the sixteenth century to make a house. Most comprise large logs of *Agathoxylon*, which fell into channels, were transported downstream and rapidly buried before the wood could decay fully. Modern erosion has exposed these fossil trees, which now lie horizontally on the surface of the desert. The bright and varied colours of the permineralized trees are the result of variations in the iron and trace element contents of the silica.

Permineralized wood belonging to the conifers *Protocupressinoxylon* and *Agathoxylon* is found in the Purbeck Formation of Dorset. Although this wood is also silicified like that from the Petrified Forest National Park, the trees are mostly preserved where they grew in the soil of a Late Jurassic forest. The upright stumps of the trees survived following inundation of the forest by saline water. These stumps were subsequently encased by the growth of ring-shaped 'burrs' of stromatolite in a tidal flat

RIGHT Polished section of a Late Triassic conifer tree *Agathoxylon* measuring 25 cm (10 in) in diameter from the Petrified Forest National Park in Arizona, USA.

environment. Sometimes the wood was not silicified, leaving only a hollow burr to indicate the former presence of a tree. Growth rings in the permineralized wood are very narrow and can be used to estimate a typical age of about 200 years for the *Protocupressinoxylon* trees.

Cones, leaves and wood thought to be from *Araucaria*, a genus represented today by the Monkey Puzzle Tree and Norfolk Island Pine, can be plentiful in fossil floras of the Mesozoic era. Jet is a distinctive type of fossilized wood from extinct gymnosperm trees, possibly araucarians. This intensely black and shiny gemstone with a conchoidal fracture was greatly favoured by Queen Victoria during her long period of mourning the death of her husband Prince Albert. The process of jet formation – jetification – is not well understood and may have been complex, involving wood drifting out to sea, becoming waterlogged, sinking to the seafloor, being buried in anoxic mud retarding its decomposition, then re-exposed to the atmosphere, buried again and finally compacted by the ever-increasing weight of overlying sediments. Although jet is best known from Early Jurassic marine deposits in the vicinity of Whitby in North Yorkshire, England it also occurs in northern Spain. Jet is an exceptional kind of fossil wood found in marine rocks – most marine wood is very different, crumbly and cannot be carved like jet.

Cycads and bennettitales (formerly known as cycadeoids) are superficially similar groups of palm-like plants typical of the Mesozoic. The oldest certain examples of both

RIGHT Stromatolite burr that grew around the trunk of a conifer tree from the Late Jurassic to Early Cretaceous Purbeck Formation of Dorset, England. This example is now a decorative feature of a garden in Regents Park, London, England.

groups are recorded from the Permian, but whereas cycads are extant, bennettitales became extinct near the end of the Cretaceous period. The leaves of the two groups can be almost indistinguishable – varying from simple and blade-shaped to compound and pinnate – but bennettitales possessed flower-like reproductive organs, which are lacking in cycads. In both groups the leaves are generally borne in a spiral pattern around an unbranched trunk, older leaves becoming detached with growth, leaving scars on the lower parts of the trunk.

Seed ferns (pteridosperms), mentioned earlier because of their occurrence in Carboniferous coal forests, were important components of Mesozoic floras such as that found in the Middle Jurassic Cloughton Formation of North Yorkshire. More than 250 species of plants have been described from this formation, among which is the seed fern *Caytonia nathorstii*, named for its occurrence at Cayton Bay south of Scarborough. *Caytonia* probably grew as small trees bearing compound leaves, each with four long leaflets, and small, fleshy seeds. Because some palaeobotanists believe *Caytonia* to be very closely related to angiosperms, fossil *Caytonia* might be important in understanding how flowering plants evolved. The Cloughton Formation also contains fern, horsetail and *Ginkgo* fossils, beds full of vertical plant rootlets and occasional thin bands of coal. These include large fronds of the royal fern genus *Cladophlebis* which lived in the understorey of the forest, and the distinctive leaves of the maidenhair tree *Ginkgo*, an extant gymnosperm so dissimilar to

LEFT Leaves of the bennettitalean *Zamites gigas* from the Middle Jurassic Cloughton Formation of North Yorkshire, England adorn a piece of rock 40 cm (15¾ in) in height.

others that it is classified in its own division (Ginkgophyta). Inhabiting marshy environments, the horsetail *Equisetum* has been described as by far the commonest plant in the Cloughton Formation, and is the only plant to be preserved where it grew rather than being transported to the site of burial. Stems of *Equisetum* are found upright in densities reaching 20 per square metre, each stem up to 50 cm (19¾ in) tall and 4–5 cm (1½–2 in) wide near the base where it bends at right angles into the horizontal underground rhizome.

Angiosperms

With more than 300,000 living species, angiosperms – flowering plants – massively dominate modern floras. The great majority of garden and agricultural plants are angiosperms, as are all grasses and most trees. While the fossil record of angiosperms is no worse than that of other plants, there are clear gaps, which have allowed debate about exactly when angiosperms originated. Darwin famously wrote in 1879 that the origin of angiosperms was 'an abominable mystery'. More recently, molecular clocks have been used to claim an earlier date of origin for angiosperms than a literal reading of the fossil record would suggest. It is certainly possible – even likely – that the earliest angiosperms were small and never formed woody trees, or that they inhabited upland areas less well represented by sedimentary rocks. The rise of angiosperms had profound consequences

RIGHT Fossils of carbonized plant roots penetrating into sediment of the Middle Jurassic Cloughton Formation at an outcrop on the Yorkshire Coast, England. The field of view is about 17 cm (6½ in) across.

for terrestrial life, triggering substantial increases in diversity among insects and other animal groups, especially those inhabiting tropical rain forests.

Among the oldest fossil angiosperm flowers, from the Early Cretaceous, are tiny examples preserved as fusain, which is indicative of burning. By the end of the Cretaceous, angiosperms were the most diverse group of plants, and woody angiosperm trees had begun to appear in numbers alongside gymnosperm trees. Fossilized wood from these two groups can be difficult to tell apart, although angiosperm wood typically contains vessels and fibres that are not found in gymnosperm wood. As for the flowers that define angiosperms, exquisite examples have been found in Eocene Baltic amber. Most are small, measuring less than 10 mm (½ in), but the largest known is a *Symplocos* flower that is 28 mm (1 in) in diameter and contains pollen.

Until quite recently it was believed that grasses, such a ubiquitous element of modern vegetation, did not evolve until after the Cretaceous. However, the discovery of distinctive grass phytoliths in dinosaur coprolites of Late Cretaceous age proved otherwise. Phytoliths are tiny siliceous structures that are found in the tissues of many plants. They probably perform supportive and anti-predator roles. Even though the fossil record of grasses has now been pushed back into the time of the dinosaurs, it may have been 50 million years later, in the Oligocene, before extensive grasslands appeared on our planet.

The London Clay – as the name suggests – lies under much of London but also parts of Sussex and Hampshire to the west, and Kent and Essex to the east. It is a formation globally famous for its rich fossil flora. This flora has been known for more than 350 years ever since writer and horticulturist John Evelyn (1620–1706) first mentioned London Clay plant fossils in 1668. It is dominated by the seeds and fruits of angiosperms belonging to 300 different species, which were described in great detail by Eleanor Reid and Marjorie Chandler in their book *The London Clay Flora* of 1933. Most numerous are fruits of the mangrove palm *Nypa*. Today, this plant genus occurs in coastal regions of Southeast Asia, and its Eocene forerunner also lived at the edges of the land areas bordering the shallow sea in which the London Clay was deposited. Also present are species belonging to the laurel, custard apple and frankincense families, along with miscellaneous wood and twig fossils. Taken as a whole, the London Clay flora points to the presence of a paratropical rainforest over southern Britain 50 million years ago. London Clay fruits are often preserved as pyrite permineralizations, their weight causing them to become concentrated along the drift lines of present-day coastal localities such as Sheppey in Kent. Unfortunately, these pyritized fossils tend to oxidize and disintegrate when exposed to the atmosphere, requiring particular measures for conservation.

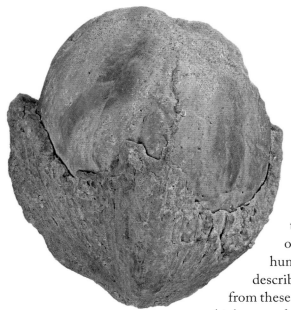

LEFT The fruit of *Nypa burtinii*, a mangrove palm. This specimen from the Eocene of Schaerbeek in Belgium is 11 cm (4¼ in) wide.

Some of the most beautiful fossil angiosperms in the collections of the Natural History Museum came from a Miocene site at Oeningen (or Oehingen) in Baden-Wurttemberg, Germany. The famous geologist Charles Lyell wrote at length about the Oeningen fossils, mentioning not only the plants but also a supposed human skeleton ('*Homo diluvii testis*') described in the early eighteenth century from these deposits by Johann Jakob Scheuchzer, which turned out to be a large amphibian. The sediments at Oeningen were deposited in a lake into which fell the leaves, seeds and even flowers of trees growing around the lake, accompanied by insects and spiders. Preserved as compression fossils, the Oeningen angiosperms belong to extant genera and include alder, sweet chestnut, oak, walnut, aspen, willows, elm, hawthorn and maple trees. In short, this 12-million-year-old flora had a distinctly modern aspect.

Marine plants

Although we find an abundance of seaweeds on modern rocky shores, the fossil record contains rather few convincing examples. Brown algae (Phaeophyceae), such as the seaweeds *Fucus* and *Laminaria*, not only lack hard parts but they usually inhabit places where erosion rather than deposition is occurring, further reducing their chances of being fossilized. Other types of marine algae, however, do have mineralized skeletons and these can be common, though they generally make unspectacular fossils.

Calcareous green algae (Chlorophyta) living today include *Halimeda*, common in shallow waters of the tropics, and while a modern red alga (Rhodophyta) called *Corallina* that is found globally also secretes a resilient skeleton of calcium carbonate. Numerous fossils, some present in rock-forming quantities, have been identified as belonging to one or other of these groups but many lack the diagnostic features

LEFT Remarkably preserved delicate flower of *Chaneya oeningenensis*, measuring 2 cm (¾ in) across, from the famous Miocene fossil locality at Oeningen on Germany.

MIDDLE LEFT Fossil leaves from the Oeningen Miocene. The large leaf on the right is from a maple, *Acer trilobatum*, and measures about 10 cm (4 in) across.

BOTTOM LEFT This angiosperm fossil, also from the Miocene of Oeningen, is of a winged maple seed, 9 cm (3½ in) in length.

needed to pin down their identities with certainty and thin sections are required to appreciate their structure. One of the earliest is the probable green alga *Koninckopora*, a tubular fossil covered by a grid of small hexagonal pores, which is found from the Early Carboniferous limestones of Britain and continental Europe.

Although the oldest red alga (Rhodophyta) may be the Precambrian genus *Bangiomorpha* (p. 35), fossils of 'coralline' red algae with calcareous biomineralization date back to at least the Silurian and are found commonly in Cretaceous and Cenozoic rocks. Crust-like corallines are important in building reefs, whereas unattached corallines form nodules called rhodoliths. These range in shape from lumpy to almost perfectly spherical. Large areas of the modern seafloor are littered with rhodoliths, and beds packed with fossil rhodoliths occur frequently in the geological record, often to be seen in building stones. Although a single coralline algal species can create a rhodolith, many rhodoliths comprise several intergrown species, often accompanied by worms and barnacles. Growth begins on a substrate, often a shell, and extends to cover it on all sides, a pattern of development only made possible by currents or animals periodically rolling the rhodoliths on the seafloor like tumbleweeds.

Distinctive fossils of another red alga, *Neosolenopora jurassica*, can be found in British Jurassic limestones. Small reefs in the Late Jurassic Portland Stone of Dorset were constructed by this alga which, when broken open, reveal conspicuous growth bands believed to be annual. But the most spectacular examples of *Neosolenopora* are to be found in Gloucestershire in a Middle Jurassic limestone known as 'beetroot stone'.

RIGHT Photograph taken in the field at Pietraroja in Campania, Italy, showing several Miocene algal rhodoliths with concentric internal laminations. The pencil at the bottom left serves as a scale.

LEFT The strong red colouration of the calcareous alga *Neosolenopora jurassica* led to the name 'beetroot stone' for the Middle Jurassic limestone in Gloucestershire, England where it is found. Measuring 18 cm (7 in) across, this specimen was more than 16 years old to judge from the presumed annual growth bands.

The stone takes its name from the delicate pink colour of the *Neosolenopora* specimens, a vestige of the original red pigmentation of the living plant.

Finally, some angiosperms have evolved to live in marine environments, most notably seagrasses. Fossils of seagrasses are rare, but their former existence can be inferred from the presence of fossils of animals that lived on their leaves. For example, impressions of seagrass leaves have been found on the undersides of encrusting bryozoans in the Late Cretaceous limestones of Maastricht, the Netherlands.

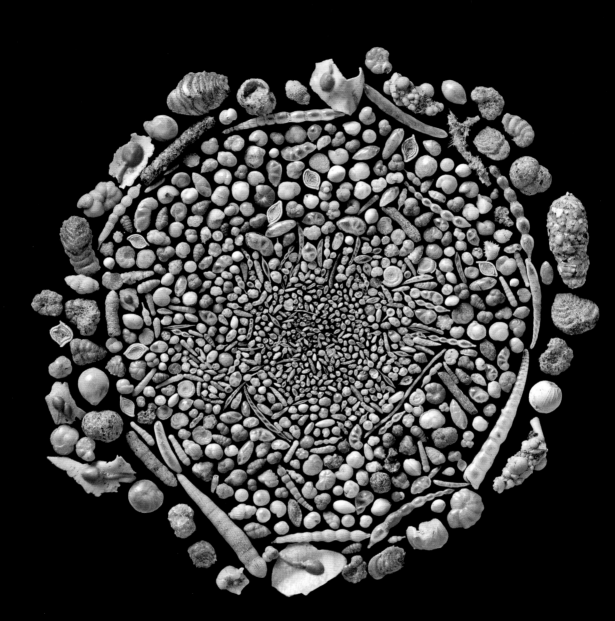

12 Microfossils by the millions

The smallest fossils are the most numerous. A piece of sedimentary rock that can be held in the palm of one hand can easily contain thousands of microfossils. But they will not necessarily be obvious because the majority of microfossils are so small that a lens or a microscope is needed to see them. Indeed, study of the tiniest microfossils, which are less than one-hundreth of a millimetre in size, requires a high-resolution microscope such as a scanning electron microscope. Some microfossils are the skeletal remains of tiny organisms (e.g. foraminifera and ostracods), but others are dissociated parts of larger organisms (e.g. conodonts) or reproductive bodies. A distinct subgroup of microfossils called palynomorphs have organic skeletons, and include spores and pollen of terrestrial plants as well as small planktonic organisms from aquatic environments.

Microfossils are of immense geological importance. Because of their minute size and high abundance, they can be found in drill cores and chippings recovered from boreholes or oil wells. They are useful not only in the stratigraphical correlation of rocks (p. 22), but also for tracking palaeoenvironmental changes at very fine timescales. Modern instruments capable of analysing the isotopic compositions of the tiny skeletons of microfossils have enabled huge advances in our understanding of climate change through geological time. Research on microfossils found in cores drilled into the thin but pristine covering of sediments beneath the ocean floor has revolutionized our knowledge of ancient climates while also allowing the evolution of these tiny organisms to be followed in great detail (see p. 224).

Most microfossils are too small to be collected individually in the field. Instead, bulk samples of sediment must be collected, broken down and washed using a sieve. The microfossils can then be picked from the clean residue using a fine paintbrush wetted with water to which the individual microfossils will adhere for transfer to a cavity slide.

OPPOSITE Decorative arrangement of foraminifera of different shapes and sizes collected during the oceanographic expedition of HMS *Challenger* during 1873–6. Similar foraminifera are common as microfossils.

> ## FOSSILS AND EVOLUTION
>
> Everyone knows that evolution – whether biological or cultural – means gradual change through time. But biological evolution is more complicated than just this. Biological evolution has at its centre the fact that the kinds of animals and plants inhabiting our planet have changed over time. Clear evidence for this comes from the striking differences between the fossils found in rocks of different ages. Fossils show, for example, that Palaeozoic seas teemed with trilobites but there were no ichthyosaurs, whereas there were no trilobites in the younger seas of the Mesozoic in which ichthyosaurs swam. For the early naturalist Georges Cuvier (1769–1832) and his followers, changes in fossils through geological time were not due to evolution but instead resulted from succession extinctions followed by new creations of life.
>
> Evolution, however, entails much more than just differences between organisms living at different times. In the strict sense, biological evolution is defined as descent with modification, i.e. individuals from one generation to the next being different. Such inherited differences – each of which may in itself be infinitesimally small – when accumulated across thousands of generations can lead to major changes and the formation of new species. The fossil records of most groups are too patchy to track descent with modification in fine detail. However, planktonic foraminifera are extremely abundant, precisely dated and have been sampled from countless sites across the globe, offering a realistic possibility for tracking evolutionary change within lineages and the origin of new lineages. A classic 2011 study of over 300 species of Cenozoic planktonic foraminiferans revealed changes in their morphologies through time. In about 40% of cases, new morphologically distinct species were found to have originated by gradual transformation of one species into another along a lineage. In the remaining 60% of cases, new species arose from lineage splitting. This was either by branching – the formation of two daughter species with extinction of the parent species – or budding – the formation of a single daughter species with continuation of the parent species.
>
> By proposing a viable mechanism – Natural Selection – Charles Darwin and Alfred Russel Wallace were able to convince many people of the reality of evolution, something that earlier proponents of evolution had failed to do. Paraphrased 'the survival of the fittest', Natural Selection hypothesizes that those individuals in a population that possess traits making them better adapted to their environment will survive better and consequently tend to leave a greater number of offspring. The proportion of individuals possessing advantageous traits will therefore increase through time. Although evolutionary processes such as Natural Selection are best studied using living organisms, fossils reveal the ever-changing unfolding of life stretching back more than 3,500 million years, which can be only reasonably explained by evolution.

Coccoliths

The tiniest microfossils of all are known as nannofossils, derived from the Greek 'nanos' meaning small. They include coccoliths, the calcareous plates of a group of planktonic plants that inhabit the well-lit surface waters of the oceans. Coccoliths are so tiny that millions would be needed to cover the surface of a disc 1 cm (½ in) in diameter. Despite their minuscule size, coccoliths occur in such prodigious quantities that they are the principal constituents of the Chalk, a limestone deposited across much of northern Europe through the 35 million years of the Late Cretaceous. Given that the Chalk can reach over 1,000 m (3,280 ft) in thickness, the total number of coccoliths present in this rock must be truly astronomical. Fossil coccoliths first appeared relatively recently in geological terms, during the Late Triassic. They are sufficiently diverse and abundant in the Cretaceous and Cenozoic to enable the establishment of a nannofossil zonal scheme founded mainly on coccolith species ranges.

But what exactly are coccoliths? Coccoliths are tiny calcite plates formed by single-celled haptophyte algae called coccolithophores. Periodic blooms of modern coccolithophores produce so much calcium carbonate that they can turn swathes of the ocean surface white, as seen in satellite images. Indeed, coccolithophores produce more calcium carbonate than any other single group of organisms on our planet, a fact not solely of academic interest given the role of calcium carbonate biomineralizers in removing the greenhouse gas carbon dioxide.

Individual coccoliths make up a coccosphere, a covering on the outside of the single cell of the coccolithophore that may serve a defensive function against predators or viral infections. Coccoliths are geometrical marvels, with an intricacy belying the apparent simplicity of the cells making them. Each is a single calcite crystal manufactured within the cell and pushed to the outside when complete. Coccoliths of different species have distinctive shapes. Perhaps the best-known are those of the living species

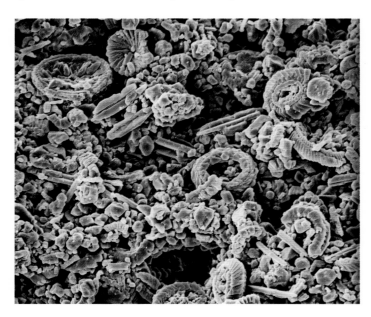

LEFT Magnified more than 2,000 times, coccoliths pack this scanning electron microscope image of a piece of Late Cretaceous Chalk from Folkestone in Kent, England.

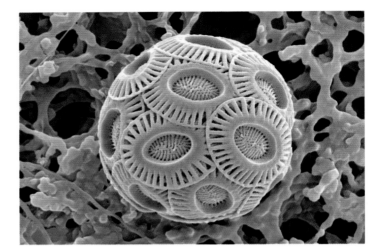

RIGHT Coccosphere of a modern specimen of *Emiliana huxleyi* measuring about 7 microns in diameter. The coccosphere is constructed of several elliptical plates of calcite called coccoliths with a distinctive spoke-like appearance.

Emiliania huxleyi, which resemble a pair of coarsely spoked wheels placed one on top of the other, the lower wheel visible through an opening in the centre of the upper wheel. Individual coccoliths are arranged in a somewhat haphazard overlapping pattern to form the protective coccosphere over the spherical cell of *Emiliania*.

Foraminifera

Anyone who has been taught basic biology at school will have heard of the amoeba, a protozoan (or protist) used as a model single-celled organism. The amoeba of the classroom has no hard parts and does not fossilize. However, its close relatives, the foraminifera – forams for short – have a rich fossil record because most either biomineralize a calcareous shell, or glue together small sediment particles to form an agglutinated skeleton. Around 40,000 species of fossil forams have been recognized, the oldest dating back to the latest Precambrian. Added to these are almost 10,000 extant species. Living forams feed on tiny particles in the surrounding water, capturing them using extensions of the cell called pseudopodia, which are also employed to dispose of waste matter. Forams are predominantly marine, and the majority are benthic, living on or within sediments of the seafloor. Some, however, are planktonic and float in the surface waters of the oceans. Numerous species of both benthic and planktonic forams host photosynthetic symbionts, providing an additional source of nutrition.

The oldest forams were entirely benthic but planktonic forams have evolved on at least four occasions from benthic ancestors, probably starting in the Jurassic period. Planktonic forams have become so numerous that their sunken shells may form a deep-sea sediment called '*Globigerina* ooze'. The sheer abundance of planktonic

LEFT Foraminifera in a sample of *Globigerina* ooze from the bottom of the Pacific Ocean. This scanning electron microscope image measures 1.1 mm across.

forams in Cretaceous and Cenozoic deposits has made them invaluable not only as zonal fossils but also for portraying evolution in deep time (see p. 224).

Chambered shells, or tests as they are more often known, are a defining feature of forams. During growth, new chambers are added at one end of the test. The chambers are assembled in straight or zig-zag lines in some species, though in most they are arranged spirally and at first glance may resemble tiny ammonites. Indeed, forams were believed to be small ammonites by some early nineteenth century naturalists. In many species the test is a helicospiral coil, as in gastropods, and it is possible to distinguish left- and right-handed varieties. Some species contain a mixture of left- and right-handed individuals. In living forams belonging to the genus *Ammonia*, left-handers become more common as seawater temperature rises, and the ratio of left- to right-handed fossil tests in various fossil forams has been used to infer seawater temperatures in the geological past.

Foram tests are typically less than a millimetre in size. However, several groups from the Late Silurian onwards evolved far bigger tests of macrofossil proportions, the largest attaining 19 cm (7½ in) in diameter. These 'larger benthic forams', abbreviated LBF, have more complex tests than other forams, which in some species are constructed of multiple intertwined spiral rows of chambers. LBF are invariably associated with algal (or diatom) symbionts. They usually inhabited shallow-water tropical and subtropical seas, languishing on the seabed in the sunlit conditions essential for their symbionts to photosynthesize.

Among forams, LBF are the most likely to be noticed by the casual observer, not just owing to their large sizes but also because they are major components of some limestones used as building stones. For example, a group of LBF called fusulines on account of their spindle-shaped tests are prolific fossils in the Permian Cottonwood

ABOVE Rock 7.5 cm (3 in) across containing the large foraminiferan *Nummulites gizehensis* and three isolated tests from the Eocene of Gizeh in Egypt.

Limestone and can be seen in historic buildings in Kansas. The largest examples of LBF are found among the nummulites, coin-shaped fossils present in great quantities in Eocene rocks. Much of Girona Cathedral, Catalonia, Spain is made of nummulitic limestone, as are the pyramids of Giza in Egypt. Two names in fossil folklore have been used for nummulites from the pyramids – slaves' lentils and angels' money. Both refer to a species called *Nummulites gizehensis*, which exists in two size classes, a smaller A-form and a larger B-form. The life cycles of forams in general often entail an alternation of asexually and sexually produced generations. In the case of *Nummulites*, the A-form 'slaves' lentils' were products of asexual reproduction and the B-form 'angels' money' were the result of sexual reproduction.

Radiolaria and diatoms

Exquisite skeletons made of silica are trademarks of two groups found commonly as microfossils: radiolarians, which are a type of planktonic marine protozoan, and diatoms, which are unicellular algae found in the sea, freshwaters and even soils.

Radiolarians typically measure 0.1–0.2 mm in diameter and have a single cell equipped with thread-like extensions used to capture small food particles. Their intricate skeletons consist of radial spines, often spiked, protruding from and through a lattice-like capsule. The exact shape of these close relatives of forams varies according to species. Many are roughly spherical, others conical, spindle- or bell-shaped. The beauty of radiolarians is nowhere better seen than in Ernst Haeckel's illustrations for his 1904 book *Art forms in nature*. Perhaps as many as 1000 species of radiolarians live in the sea today, with at least 10 times that number known in a fossil record that extends back to the Cambrian. Large quantities of radiolarian skeletons are present in modern sediments accumulating below about 5,000 m (16,404 ft), a depth beyond which nearly all calcareous plankton is lost to dissolution. These 'radiolarian oozes' are particularly characteristic of the equatorial Pacific in areas where upwelling makes the surface waters highly productive. Ancient radiolarian oozes are preserved in the rock record as radiolarian cherts, which are sometimes found associated with pillow lavas, as in the Ordovician rocks of the Ballantrae area of southern Scotland.

Setting aside some Jurassic fossils of doubtful identity, diatoms made a very recent debut in the fossil record, the oldest unequivocal examples coming from the Cretaceous. Yet they have more than made up for their late appearance by diversifying into an astonishing 200,000 species. It has been estimated that diatoms today constitute almost half of the living organic biomass in our oceans. Like radiolarians, diatoms are marvels of microengineering. The diatom cell wall is impregnated with silica to form two filigree valves, one fitting over the other like a minute pillbox. This

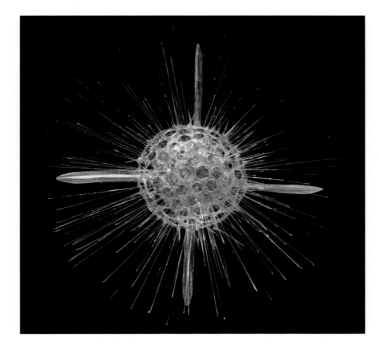

LEFT Glass model of the radiolarian *Actinomma asteracanthion* made by Leopold and Rudolf Blaschka during the late 19th century. The central capsule is about 90 microns in diameter in living specimens of this protist.

skeleton, known as a frustule, ranges from about 0.002–0.2 mm in size and is usually either circular or elliptical in outline shape, although some diatoms have triangular frustules. In a few species the frustules are stacked to form a colony. As diatoms are photosynthetic organisms, they need to inhabit well-illuminated environments, some floating in the surface waters of seas or lakes, others settling on the surfaces of marine plants, shells and rocks. Vast amounts of siliceous sediment are generated by diatoms. A soft rock called diatomite results when diatom frustules overwhelm other sediment particles. Formed in the sea or lakes, diatomites are commercially exploited for many purposes, including the manufacture of filtration systems. A quarried diatomite at Foulden Maar in the South Island of New Zealand consists of the frustules of diatoms that flourished in a Miocene lake occupying the crater of an extinct volcano. The finely laminated Foulden Maar diatomite contains a spectacular array of fossil fishes, insects and plants, in addition to the diatoms themselves.

RIGHT Scanning electron microscope image of fossil radiolarians from the Cenozoic of Barbados. The field of view is about 1.8 mm across.

BELOW RIGHT *Encyonema jordaniforme* is one of the main diatom species forming a Miocene diatomite at Foulden Maar in Otago, New Zealand. Frustules of this unicellular alga are seen in a scanning electron microscope image which has a field of view 65 microns across.

Ostracods

Crabs, prawns and lobsters (see Chapter 3) are familiar types of crustaceans, but there are other groups belonging to this subphylum of arthropods that are diverse but unknown to most people. Among these are crustaceans classified in the class Ostracoda, the ostracods or 'seed shrimps'. There are estimated to be more than 65,000 different species of fossil and living ostracods. Their fossil record extends back to the Early Ordovician, discounting some ostracod-like Cambrian fossils, and they can be found living today in marine and freshwater environments, including lakes. A few even inhabit damp soils. Feeding mode varies according to species. Some are filter feeders, others feed on detritus or scavenge, and ostracods can be herbivores or carnivores.

Averaging about a millimetre in size, ostracods clearly qualify as microfossils, even though a few exceptional fossil species reached 25 mm (1 in) across. They can be recognized by their carapaces or shells, which consist of two hinged valves of calcite resembling a tiny bivalve mollusc. However, unlike bivalve shells, those of ostracods lack growth lines because ostracods grow by moulting their old shells and manufacturing larger ones to replace them instead of by adding new increments to the existing shells. Up to eight moult stages can occur during the lifetime of an individual ostracod. Bean- or kidney-shaped, oval or elongated in shape, ostracod shells range from smooth and relatively featureless, to ornamented by a variety of ribs, spines and tubercles, some having a reticulated or pitted surface. Tiny scars may be visible on interior surfaces of shells marking points of attachment for the adductor muscles used to close the shell as well as other types of muscles. The valves are almost but not quite mirror images, one being slightly larger than the other, which it overlaps.

The soft body of ostracods comprises a head and thorax and up to eight paired appendages including antennae, mandibles and segmented walking legs. The shell is all that remains in the great majority of fossil ostracods. There are, however, some notable examples of soft part preservation resulting from replacement or coating by phosphate minerals, silica, pyrite or calcite. For example, concretions from the Cretaceous Santana Formation of Brazil have yielded more than 700 phosphatized specimens of *Pattersoncypris micropapillosa*. Limbs, antennae, mandibles and mouths are among the soft parts preserved, likely made possible by the release of phosphate from the carcasses of decaying fish, whose fossils are found in the same concretions.

Because different species live in a wide variety of aquatic environments of varying salinity, ostracods have proved very useful in research on palaeoenvironments. For instance, the shells of freshwater species tend to be smooth, weakly mineralized and bean-shaped. At the other end of the spectrum are ostracods inhabiting high-energy marine environments; these often have thick shells with reticulations or strong ribs. The ostracod

 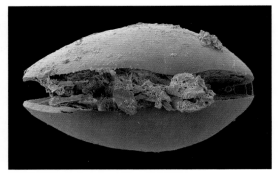

ABOVE Scanning electron microscope image of an artificial cast of the Late Silurian ostracod *Calcaribeyrichia* from Radnor Wood in Shropshire, England. The fossil is about 1 mm in size.

ABOVE RIGHT This specimen of the Brazilian Cretaceous ostracod *Pattersoncypris micropapillosa* measuring about 1 mm in size is exceptional in preserving not only the shell but also phospatized limbs.

genus *Cyprideis* always lives in shallow waters. When drilling through sedimentary rocks at the bottom of the Mediterranean Sea, the discovery of a layer containing *Cyprideis* within otherwise deep-sea sediments formed part of the evidence for the drying-up of the Mediterranean during the late Miocene. The chemical composition of ostracod shells can also provide valuable information on the temperature, salinity and biological productivity of ancient environments. Following moulting, the rapid biomineralization of new shells uses chemicals taken from the surrounding water. This means that the elements and isotopes in the ostracod shell reflect the chemistry of the water at that moment of time.

Conodonts

In 1856 the biologist Heinz Christian Pander (1794–1865) coined the name 'conodonts' for some unusual tooth-like microfossils found in Palaeozoic rocks which he believed to be fish teeth. Individual 'conodont elements', as they are generally called, usually range from 0.2 to 2 mm in size and are made of the phosphate mineral hydroxylapatite. Unlike microfossils with soluble skeletons of calcium carbonate, conodont elements can be extracted by soaking limestones in 10% acetic acid and sieving and examining the insoluble residue. A kilogramme of rock may yield upwards of a thousand conodont elements. Three main shapes of conodont elements are recognized. The simplest – cones – have a single curved, horn-shaped cusp; next come blades, which have a main cusp plus several smaller cusps (denticles) on each side; finally, there are platform elements in which a ridge bearing numerous cusps of similar size arises from an expanded base.

The oldest unequivocal conodonts come from rocks of Late Cambrian age, although some fossils resembling conodonts have been recorded in older rocks close to the Cambrian–Precambrian boundary. Long thought to have become extinct near the end of the Triassic period, recent discoveries of Jurassic conodonts in Hungary and Japan show that they survived for a little longer. Conodonts have been extensively used for stratigraphical correlation. Indeed, they are the pre-eminent microfossil group for this purpose in rocks of Ordovician–Permian age, and a conodont zonation of the Triassic has also been devised. For instance, 40 conodont zones are proposed in rocks deposited during the 53-million-year duration of the Permian period, each zone representing on average slightly over 1 million years of time.

For many years palaeontologists were perplexed by conodonts: what kind of animal did they come from, and what was the function of these tooth-like elements that, unlike normal teeth, apparently never showed surface wear? Perhaps they were used in filter feeding? Several fanciful reconstructions of the conodont animals were proposed, none altogether convincing. The foundation for most of these reconstructions were rare finds of groups of associated elements. Found on bedding planes, these 'conodont assemblages' consist of 6–11 pairs of conodont elements of various kinds arranged symmetrically on either side of a long axis. Several different conodont genera, including blades and platforms, are present in each assemblage, meaning that the genera themselves are artificial in the sense of not corresponding to a single biological genus.

The discovery in 1982 of a soft-bodied fossil containing conodont elements provided the answer to the conodont conundrum. This fossil was found in the Early Carboniferous Granton Shrimp Bed on the southern shore of the Firth of Forth in Scotland. Recognized as *Clydagnathus* based on element morphology, the compressed, eel-like body of the animal is about 4 cm (1½ in) long and has an assemblage of conodont elements at the head end. Chevron structures running along the body are interpreted as muscle blocks

LEFT Conodont elements from the Silurian of Gotland, Sweden. The field of view in this scanning electron microscope image is 5 mm across.

of the kind typical of vertebrates. Other conodonts with soft parts preserved have since been found elsewhere, notably an Ordovician species from the Soom Shale of South Africa, which was a much larger animal at least 40 cm (15¾ in) long. The consensus among palaeontologists is that conodonts formed a branch of the vertebrate tree just above the extant hagfishes and lampreys. If this is correct, conodont elements represent the earliest known biomineralized structures to have evolved among vertebrates.

As knowledge of the anatomy and affinity of conodonts has developed, it has become easier to deduce their palaeoecology. Conodonts swam, often close to the seafloor, guided by a pair of large eyes, which are evident in some of the soft-bodied fossils. They were predators or scavengers employing the various kinds of elements around the mouth to capture and process food. Earlier claims that conodont elements never show signs of wear have been repudiated by observations of extensive wear and damage incurred while functioning as teeth.

Palynomorphs

Palynomorphs are a distinct subset of microfossils. These organic microfossils are generally found among the insoluble residues when clastic sedimentary rocks are dissolved in highly corrosive hydrofluoric acid. They fall into two main groups: the aquatic dinoflagellates and acritarchs, and the mostly terrestrial spores, cryptospores and pollen which, however, are often transported and deposited in sediments accumulating at the bottoms of freshwater lakes.

Dinoflagellates are single-celled algae between 20 and 150 microns in size. The majority of the 2,300 or so extant species live in the sea with a few inhabiting freshwater environments. Their name is derived from the presence of two whip-like flagella used to aid propulsion in the plankton. Some living dinoflagellates are bioluminescent, imparting a blue-green colour to the sea at night. The fossil record of dinoflagellates is mainly due to the cysts – dinocysts – made by a minority of species during a sexual stage in their life cycles. Dinocysts are benthic, lying dormant on or in the sediment until favourable conditions return for the motile, planktonic stages to flourish, sometimes initiating blooms such as the toxic 'red tides' for which dinoflagellates are notorious. Most dinocysts possess extremely resistant walls of an organic compound called dinosporin. The oldest fossil dinocysts come from the Triassic. By the Jurassic they were common enough to allow a system of zones and marker horizons for stratigraphical correlation to be devised. Dinocysts vary greatly in morphology. Many are spherical and covered by spines or similar processes, whereas others have a plated surface more closely resembling the motile stages of the dinoflagellate species that produced them.

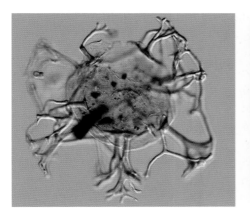 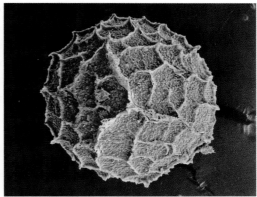

ABOVE Jurassic dinoflagellate *Rigaudella aemula*. This example from Dorset, England is about 80 microns across.

ABOVE RIGHT Female spore of the living clubmoss *Selaginella kraussiana* about 300 microns across. Very similar spores have been found as fossils at least as far back as the Cretaceous period.

Whereas dinoflagellates are a true biological group, constituting the superclass Dinoflagellata, another group of aquatic palynomorphs called acritarchs are a taxonomic wastebasket containing cysts of unknown origin. Unsurprisingly, acritarchs have a long geological range. The oldest are over 1500 million years old but a large number of named genera are of early Palaeozoic age, diversifying in concert with macrofossils during the Cambrian Explosion (p. 39).

Many land plants produce spores for reproduction and dispersal. Because they are made of a highly resistant polymer called sporopollenin, spores fossilize extremely well. Ordovician–Devonian spores of the oldest land plant were named 'cryptospores', meaning hidden spores, because they pre-date macrofossils of terrestrial plants in the fossil record (see Chapter 11). Pollen first appeared in the Mesozoic. This is a type of male microspore usually associated with flowering plants, although the most ancient pollen from the Late Devonian was probably made by gymnosperms. Analysis of the pollen present in Quaternary lake and peatbog sediments is a routine way of tracking changes in vegetation through time; for example, a switchover from dominantly grass pollen to tree pollen would signal the transition from grassland to woodland. Turning the focus back 66 million years to the time of the mass extinction at the end of the Cretaceous period, research on spores and pollen has revealed a sharp increase in the abundance of fern spores in sites as widely separated as North America and New Zealand. This 'fern spike' is interpreted as resulting from the proliferation of ferns following on from destruction of woodland vegetation by wildfires and other forms of environmental disturbance after the famous asteroid strike (p. 26).

13 Trace fossils: snapshots of ancient behaviour

Those who have walked along a beach and looked back across the damp sand will see a transient record of their journey. It provides evidence of our presence, and we are not the only ones to make such prints. Animals walking across wet sand or digging into the mud at the bottom of a lake also leave tell-tale marks in the sediments. Tracks and burrows have the potential to survive into the rock record, falling into a special category of fossils called trace fossils. Trace fossils differ from body fossils in that they do not consist of the bodily parts of animals or plants, but rather the relics of their activities in or on sediments.

The study of trace fossils is called ichnology from the Greek *ichnos*, meaning footprint or track. During the nineteenth century before their origin was understood, the prevailing view was that many trace fossils were fossilized seaweeds, leading to the moniker 'fucoid markings', a name derived from the brown alga *Fucus*. Trace fossils, or ichnofossils, have their own system of naming to differentiate them from biological taxa. This is needed because the exact identity of the trace maker is seldom known, and many animals are capable of producing very similar traces. For example, observations of modern animals show that burrows of almost identical appearance can be excavated by a wide range of different animals, from worms, to crustaceans, to vertebrates. Conversely, a single animal may be capable of producing a variety of different ichnotaxa. Rather than being diagnostic of a single biological taxon, ichnotaxa reflect patterns of behaviour. There are ichnospecies and ichnofamilies, but the most commonly used ichnotaxon is the ichnogenus. This is often named using the suffix *-ichnus*, as in *Lobichnus* and *Imbrichnus* for lobate and imbricated traces, respectively.

Trace fossils attract interest from both sedimentologists and palaeontologists. They produce structures in sedimentary rocks supplementing those formed by the action of physical forces such as wind and water currents. Unlike body fossils, which can become deposited far away from where the organisms lived, trace fossils

OPPOSITE Meandering trace fossil attributed to the ichnogenus *Olivellites*. Measuring 15 cm (6 in) in height, this fossil trail is believed to be of Carboniferous age. It was probably made by an animal gathering food particles from the upper layers of sediment.

are generally preserved in situ. Consequently, they are valuable as indicators of the palaeoenvironmental conditions that existed at the site of deposition. Following the principle of uniformitarianism, studies of incipient trace fossils being produced by living animals in known environments inform us about the likely environments of similar trace fossils from the geological past.

There are many kinds of trace fossils, and a variety of ways to classify them. One popular scheme is based on the reason for the behaviour producing the trace fossil in question: was it to construct a home, to extract food items from the sediment, or to escape from predators? However, purpose is not always obvious and may be multifarious, which is why trace fossils are here treated within the simple categories of burrows, borings and scrapings, footprints, tracks and trails, and coprolites.

Burrows

One thing that earthworms, desert tortoises and moles have in common is that they all burrow. Burrowing into sediment is a routine and essential activity for many kinds of animals whether inhabiting the sea, freshwater environments or living on the land. Accordingly, fossil burrows are common in sedimentary rocks, including porous sandstones in which they may be the only fossils present following the frequent loss of body fossils to dissolution by groundwater. Sometimes individual burrows are well defined, but on other occasions they are diffuse, particularly when multiple phases of burrowing have churned up or 'bioturbated' the sediment.

Burrow shapes vary widely, as reflected in the plethora of ichnogenera coined for burrows. The majority of burrows are essentially cylindrical structures but they can be straight, curved, U-shaped, or spiral in form, simple and undivided or complexly branched, and oriented parallel or perpendicular to bedding. Some possess a distinct lining (often of sediment or faecal pellets), while others are unlined. Many burrows are maintained as open passageways, but others are back-filled by the burrowing animals using sediment packed in a meniscus-like pattern, or alternatively become infilled passively by sediment entering the burrow after it was vacated. Burrows can retain a constant diameter or vary in diameter along their lengths, as when burrows end in enlarged chambers, and may have sides that are smooth or sculpted by scratch marks (bioglyphs) created by the burrowing animal. Some burrows such as *Ophiomorpha* have walls constructed of faecal pellets or shell fragments.

One of the oldest and most simple burrows is *Skolithos*, consisting of straight tubes, up to 5 cm (2 in) in diameter and 35 cm (13¾ in) long, oriented perpendicular to the bedding. Cambrian sandstones in northwest Scotland packed with this trace fossil are known colloquially as 'pipe rock'. Although the identity of the pipe rock tracemaker is unclear, it was almost certainly a type of worm living as an ecological pioneer in shallow-water sands.

More complex and larger than *Skolithos*, *Thalassinoides* is a three-dimensional network of bifurcating and coalescing branches. Some modern decapod crustaceans, notably the ghost shrimp *Callianassa*, excavate dwelling burrows which, if fossilized, would be identified as *Thalassinoides*. However, decapods cannot be responsible for all fossil examples of this ichnogenus as body fossils of decapods do not appear until the Late Devonian whereas *Thalassinoides* is recorded back to the Cambrian. Spectacular examples of *Thalassinoides* can be seen in blocks of Jurassic limestone on the coast near Weymouth in Dorset and Filey in Yorkshire. Because the sediment infilling these burrows is more resistant to weathering, the burrows stand out in relief above the softer surrounding rock. *Thalassinoides* is also common in the Cretaceous Chalk, with

LEFT This block of Late Jurassic limestone photographed near Filey in Yorkshire, England is covered by a system of burrows belonging to the ichnogenus *Thalassinoides*. The burrows were infilled by sediment, which became harder during lithification than the surrounding sediment.

BELOW LEFT Burrows of two diameters are infilled by sediment of a darker colour in this 21 cm (8¼ in) wide Late Cretaceous rock from Pinsdorf in Austria. The smaller burrows are identifiable as the ichnogenus *Chondrites*.

ABOVE Measuring 23 cm (9 in) across, this vortex-like trace fossil is *Zoophycos*. The specimen was collected from the Middle Jurassic rocks of Montagne de Crussol in the Ardéche region of France.

ABOVE RIGHT A Spanish specimen of the trace fossil *Paleodictyon* of unknown age. Each polygon is about 1 cm (½ in) in diameter.

many examples preserved as flint nodules that have stubby branches or horn-shaped protuberances. These 'burrow flints' were formed when flint developed around an infilled *Thalassinoides* burrow but failed to replicate faithfully the entire burrow system.

A very different burrowing ichnogenus is *Zoophycos*. This usually takes the form of a corkscrew with a central vertical shaft, a tube around the perimeter, and arc-shaped marks called spreiten in the sediment between the shaft and tube. The spreiten are the most striking component of *Zoophycos* and can reach a metre in diameter. This trace fossil was made by a deposit feeder, with the spreiten formed by the animal moving back and forth from its central shaft to gather food in the sediment around. First recorded in the Cambrian, *Zoophycos* most often characterizes deep-water palaeoenvironments. Although incipient *Zoophycos* has been observed in the depths of today's oceans, the identity of the tracemaker (or tracemakers) is an enigma, with polychaete worms figuring high up on the list of potential culprits.

Another typically deep-sea burrow – *Paleodictyon* – is equally controversial with respect to its maker, and some scientists even believe it to be a body fossil. *Paleodictyon* resembles a small honeycomb, the hexagonal units each about 1–3 cm (½–1 in) across and defined by narrow ridges of sediment. *Paleodictyon* ranges from the Cambrian to at least the Eocene, and similar traces have been discovered using a submersible at 3,500 m (11,483 ft) depth on the mid-Atlantic ridge. Regrettably, no biological tissue was found in association with these recent structures. It is commonly believed that *Paleodictyon* is a structure made by an unknown animal for the 'farming' of bacteria.

The fossil burrows mentioned so far were all formed underwater. But, of course, burrows are also made in the soil by animals such as earthworms and moles, as any gardener would attest. At the small end of the size spectrum are typically flask-shaped burrows excavated by insects in ancient soils (palaeosols) as nests. These include the ichnogenus *Celliforma*, first recorded in the Triassic and attributed to bees or their close relatives. Probable earthworm burrows have also been identified in palaeosols dating back to at least the Triassic. Some of these are packed with faecal pellets, the lumpy exteriors of the burrows corresponding to the ichnogenus *Edaphichnium*.

ABOVE Five examples of the ichnogenus *Celliforma* from the Pleistocene of Fuerteventura, Spain. These flask-shaped nests made by hymenopteran insects measure 2–3 cm (¾–1 in) in length.

Burrows made by mammals can be on a much larger scale. For example, hyenas today dig dens up to 3 m (9 ft) deep. Fossil burrows believed to have been excavated by mammals have been identified in rocks as old as the Jurassic. They include the spectacular trace fossil *Daimonelix*, aka the 'Devil's corkscrew'. This ichnogenus was first discovered in South Dakota in Oligocene-Miocene deposits and consists of a helical shaft of hardened sediment penetrating an ancient soil to a depth of up to 3 m (9 ft). The tracemaker in the case of these South Dakotan *Daimonelix*

LEFT Excavating a corkscrew-shaped burrow called *Daimonelix* dug by an early species of beaver in South Dakota, USA.

was the extinct beaver genus *Palaeocaster*. However, not all examples of the ichnogenus *Daimonelix*, which ranges back to the Permian, were produced by mammals – similar helical burrows are dug by monitor lizards living in northwest Australia, yet another example of near-identical traces being produced by more than one kind of animal.

Borings and scrapings

Borings are often confused with burrows, but whereas burrows are formed in soft substrates by displacement of sediment grains, borings penetrate hard substrates by cutting through these grains. Hard substrates that are often bored include rocks and biogenic structures such as shells, bones and wood. The purpose of boring is usually to excavate a protected dwelling, an alternative to making a shell. However, some borers mine hard substrates for nutrition, such as so-called 'shipworms', bivalve molluscs specialized for boring into wood and consuming the cellulose and lignin, and others drill through living shells to feed on the animal within. Whatever the substrate, boring results in bioerosion and is a potent destructive force, especially in shallow seas.

Boring organisms employ both mechanical and chemical means to make holes and cavities. Calcium carbonate substrates (shells, limestones) are more often bored than

RIGHT Several types of borings made by Middle Jurassic marine animals can be seen in this 10 cm (4 in) wide block of limestone from Cleeve Hill in Gloucestershire, England. The most striking are long tubular borings attributed to the ichnogenus *Trypanites*.

LEFT Natural flint cast of *Entobia*, a boring made by sponges, often into shells. In this Late Cretaceous fossil, the cavities have become infilled by flint and the shell dissolved. The specimen measures about 9 cm (3½ in) across.

are harder non-carbonate substrates such as granite or quartzite boulders, because penetration of the former can be accomplished by chemical as well as physical means. Because borings are made in resistant hard substrates, they have a high chance of surviving as trace fossils.

Palaeontologists have coined a plethora of names for boring ichnogenera. *Trypanites* is the ichnogenus used for long cylindrical borings generally oriented approximately at right angles to the surface of the substrate and resembling the burrow *Skolithos*. Like *Skolithos*, various kinds of worms may have produced *Trypanites*, most often drilling into limestone, or penetrating corals and sponges with thick skeletons. Giant examples of *Trypanites* have been described from the Mendip Hills of England, where Jurassic worms bored up to 40 cm (15¾ in) into solid Carboniferous limestone forming a rocky platform on the seabed 170 million years ago. *Trypanites* has become ever more common in the fossil record since first appearing in the Cambrian, reflecting a general increase in bioerosion through geological time.

Many fossil collectors will have been dismayed to find what appears to be a pristine shell crumble during extraction. The reason is often because the shell was seriously weakened by boring organisms, either when the animal was still alive, or after death but before burial. Damage to shells today is frequently caused by the activities of boring sponges or polychaete worms. Boring sponges like *Cliona* excavate

RIGHT *Teredolites* is a trace fossil made by bivalves boring into wood. The wood itself has rotted away in this 27 cm (10½ in) wide specimen from the Early Eocene London Clay of Kent but the borings and their calcareous linings remain.

systems of chambers and tunnels within mollusc shells, with canals connecting to the shell surface to give the sponge access to the water for suspension feeding. The ichnogenus *Entobia* can be confidently attributed to boring sponges and is very common in Cretaceous–Holocene fossil shells. *Caulostrepsis* is a U-shaped boring that is also abundant in shells and some rocks. *Caulostrepsis* today is produced by spionid polychaetes. Evidence that it sometimes excavated the shells of living hosts can be seen in the form of blisters that the host animal produced on the inside of its shell as a response to being penetrated.

Club-shaped borings in shells and rocks belong to the ichnogenus *Gastrochaenolites*. Most were made by bivalve molluscs seeking a protected habitat. It is not uncommon to find the shell of the boring bivalve still preserved at the bottom of the boring, although the shells found inside borings occasionally belong to different 'nestling' species that took advantage of the hole as a home after the borer died.

Teredolites is the ichnogenus used for bivalve borings ('shipworms') penetrating wood and tending to be longer and more sinuous than *Gastrochaenolites*. The small shells of the tracemaker may also be preserved within the boring, and the boreholes themselves often have a calcareous lining. Thus, the entire fossil may consist of the trace fossil *Teredolites*, the diagnostic body fossil shell of the boring bivalve, and the borehole lining. Driftwood that finds its way into the sea can become heavily infested with *Teredolites*. After sinking to the seafloor and following burial in sediment, the wood can decay completely to leave only a dense cluster of fossilized *Teredolites* tube-linings as evidence of its former presence.

The trace fossil *Oichnus* was produced for an entirely different reason. It is a round hole made in shells of living animals by predators, usually drilling gastropods but occasionally octopus, which are able to chip and puncture the shell surface. This ichnogenus has been recorded in Palaeozoic shells but only becomes common in the Cretaceous and Cenozoic. The high proportion of molluscs in some fossil populations containing *Oichnus* indicates that many were victims of boring predators. Numerous

ABOVE Seven shells of Pliocene bivalve molluscs, each containing a single borehole of the ichnogenus *Oichnus* indicating that they were attacked by predatory gastropods. The field of view of this photograph taken on Chatham Island, New Zealand is about 9 cm (3½ in) across.

ABOVE RIGHT The criss-cross scratch marks visible on this Jurassic oyster shell were made by echinoids scouring the surface for food, leaving the ichnogenus *Gnathichnus*. The field of view is 3.8 cm (1½ in) across.

scientific studies have used *Oichnus* to investigate predation in the geological past, tackling such questions as: has the intensity of predation increased through geological time, have predators become better at drilling prey by positioning their drillholes in optimal locations on the shells, and were some prey species favoured over others?

When examined under a microscope, samples reveal that both living and fossil shells are frequently infested with minute borings made by tiny cyanobacteria, algae or fungi. They have different ichnogenus names, depending on the exact form of the boring. These micro-organisms are a food source for certain fishes, molluscs and echinoids which scour the bored shells. Their grazing leaves behind traces that can be observed on the surfaces of well-preserved fossils. Grazing echinoids produce star-like patterns given the ichnogenus name *Gnathichnus*. Limpets and chitons create the trace fossil *Radulichnus*, characterized by shallow gouges containing parallel scratch marks, each scratch corresponding to a single tooth on the radula of the grazing mollusc.

Footprints, tracks and trails

Few fossils are more evocative of the lives of long extinct animals than footprints in the rocks. Perhaps the most celebrated are those of 3.7-million-year-old hominins discovered in 1976 by Mary Leakey at Laetoli in Tanzania, trackways clearly made by bipedal individuals walking upright. The footsteps of the Laetoli

TOP The underside of a bedding plane in this piece of sandstone preserves the trace fossil *Cruziana furcifera*, a trail probably produced by a trilobite. This specimen comes from the Early Ordovician near Salamanca in Spain and measures about 30 cm (11¾ in) across.

ABOVE The 24 m (78 ft) long trackway at Laetoli in Tanzania preserving the footprints made by a group of Pliocene hominins.

hominins were impressed into damp ash erupted from a nearby volcano and then buried – and preserved – by subsequent ashfalls. Sedimentary rocks contain many footprints, tracks and trails made by various animals, both vertebrates and invertebrates. Some were formed on land and others in the sea. The best places to find them are in areas where contemporary erosion has exposed large expanses of bedding planes, the ancient sediment surfaces over which the trace-making animals once passed.

The quality of preservation varies enormously, from faithful moulds of the anatomy of the feet, to vague depressions of equivocal origin. Factors controlling this include the grain size and consistency of the sediment. It has been shown experimentally that footprints made in sand saturated with water are poorly defined compared with those made in damp sand. And when the sediment surface is dry and hard, footprints made in it may be too shallow to be noticeable. Footprints are either 'underprints' – where the ground on which the animal moved has been compressed beneath the surface – or 'overprints' – resulting from subsequent layers of sediment filling the concavity. Both types of preservation blur the shape and alter the size of the footprint. In addition, footprints are frequently recovered as natural casts made by infilling sediment and can be visible on the undersides of bedding planes.

Back in the Ediacaran before animals with legs had appeared, invertebrates were leaving simple trails on the sediment surface.

Trails proliferated in the Cambrian and included the widespread ichnogenus *Coclichnus* used for small sinuous or meandering trails apparently made mainly by worms. Some invertebrates are responsible for fossil trackways too. The Carboniferous–Cretaceous ichnogenus *Lithographus* is a small trackway thought to have been made by insects such as cockroaches. Much larger in size is *Diplichnites*, consisting of two rows up to 36 cm (14¼ in) apart of closely spaced prints, each elongated approximately at right angles to the line of prints. Examples from the Carboniferous were almost certainly made by the giant millepede *Arthropleura* which grew to over 2 m (6½ ft) in length and had 28–32 pairs of legs. A more common probable arthropod trace fossil is *Cruziana*, which some palaeontologists regard as a surface trail, others as a shallow burrow. This ichnogenus consists of a bilobate furrow covered by transverse or herringbone-shaped ridges. Trilobites are the likely tracemakers for most examples of *Cruziana*, but they cannot have been responsible for all, as the ichnogenus has been found in Triassic rocks postdating the last trilobite body fossils. So, the source here remains a mystery.

Turning to vertebrates, fish trails are meandering or scalloped grooves forming repeat and often overlapping patterns on bedding planes of sediments deposited below water. Tetrapods may also produce trails if their tails contact the sediment surface, but these are usually less common as trace fossils than footprints and trackways. The earliest tetrapod tracks are recorded from the Devonian. Trackways of quadrupedal amphibians, reptiles and mammals have manus (forelimb) and pes (hindlimb) impressions; those of bipedal reptiles, birds and mammals have pes impressions only. The mapping of trackways can provide a huge amount of information. Large numbers of similar trackways oriented in the same direction can point to herding behaviour. It is also possible to estimate the speed of locomotion from the length of the footprints and the distance between them (i.e. stride length). Results from the fastest dinosaurs suggest running speeds of 15 metres per second – far out-pacing Usain Bolt, who managed 10.44 mps in his world-record run in 2009.

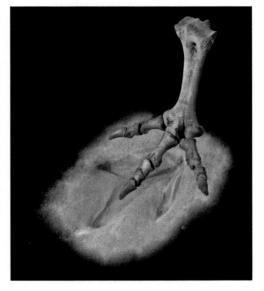

ABOVE Foot bones of a flightless moa superimposed over the trace fossil of a footprint made in now-hardened sand by a similar bird and found at Poverty Bay in New Zealand. The footprint measures about 20 cm (7¾ in) in length.

ABOVE Tridactyl footprint of a theropod dinosaur photographed at an outcrop of the Jurassic Navajo Sandstone in Utah, USA.

Discoveries of dinosaur footprints and trackways seldom fail to generate media interest, even though these are by no means uncommon. Most were formed in coastal plain environments, which were widespread and extensive in the Jurassic and Cretaceous when the edges of the continents were flooded by shallow seas. Dinosaur trackways have been known for centuries, prompting some strange early interpretations. For example, trackways observed in Portuguese cliffs were once believed by local fishermen to be the footprints of a giant mule carrying the Virgin Mary.

Despite the extreme rarity of dinosaur bones, the North Yorkshire shoreline has justifiably come to be known as the Dinosaur Coast because the Middle Jurassic rocks between Scarborough and Whitby contain a wealth of dinosaur footprints. These can be found by anyone with a keen eye, particularly in blocks of sandstone eroded from the cliffs. The clearest are tridactyl prints formed mostly by theropods; less obvious are the almost circular footprints of sauropods. No fewer than 17 different types of tridactyl footprints have been identified on the Dinosaur Coast. Some were apparently made by dinosaurs swimming in shallow water, their hind feet raking the sediment to leave elongate footprints. Persistent trampling by dinosaurs disturbed and disrupted the layering of some beds, resulting in the sedimentary structure known as 'dinoturbation'.

Coprolites

Nearly all the trace fossils mentioned are in some ways connected with feeding. The inevitable corollary of feeding is defaecation, producing excrement which, when fossilized, is known as a coprolite. This term was first coined in 1829 by William Buckland (1784–1856) who was fascinated by the coprolites found in the Jurassic rocks of Dorset, England, which he believed to be ichthyosaur faeces. Faeces do not get prettier with age – coprolites are rarely attractive and, as fossils, they are hard to classify. Numerous ichnotaxa have been established to distinguish coprolites of different shapes, but these are not universally used.

It might seem improbable that excrement could ever be fossilized but coprolites are quite abundant in some places. Carnivore coprolites are commoner than herbivore coprolites because herbivores defaecate a lot of undigested plant material attractive to scavengers. Furthermore, carnivore coprolites contain a high phosphate content due to the consumption of bones and teeth by carnivores. The phosphate forms the basis for the growth of phosphate minerals facilitated by bacterial action, cementing and hardening the coprolite. Because phosphate-rich coprolites are extremely durable, they are prone to reworking from older sediments and, unlike most other trace fossils, may be transported some distance from where the animal defaecated. Coprolites often become concentrated in condensed deposits, such as bone beds, formed during times of very slow sedimentation on the seafloor. An indication of the local abundance of coprolites and associated bones and phosphatic concretions is that they were once quarried as a source of agricultural phosphate during the nineteenth century in East Anglia, including from Plio-Pleistocene rocks near Ipswich, a town which boasts a road called Coprolite Street reflecting the former economic importance of these fossil faeces.

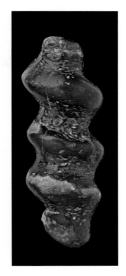

ABOVE This 10 cm (4 in) long spiral coprolite from East Anglia, England is the fossilized faeces of a shark and has been given the ichnospecies name *Helicocoprus clarki*.

Coprolites have been recorded back to the Cambrian, increasing in abundance and variety with the rise of vertebrate animals. The presence in many coprolites of food remains – including bones, teeth, hairs, insect cuticles, parasites and plants – allows them to be used for investigating diets and reconstructing wood webs in ancient communities. However, the detailed dietary evidence that coprolites can provide is constrained by the fact that the identity of the producer is seldom known for certain. Even the more distinctive coprolites with striking helical shapes moulded by digested food passing through a spiral intestinal valve can be produced by several types of fish including sharks and lungfish. A recent study of Triassic coprolites from Silesia revealed the remains of beetles, showing that these insects were consumed by the coprolite-producing animal. Although the finger of suspicion pointed to the small dinosaur-like reptile *Silesaurus*, body fossils of which were found in the same locality, finding proof for this link proved impossible.

Whereas most research has focused on vertebrate coprolites, with dinosaur poo attracting particular attention, clusters of small pellet-like 'microcoprolites' made by invertebrates can also be fossilized, particularly in the protected settings afforded by burrows, borings and shell interiors.

Further information

Fortey, Richard A. *Life: An Unauthorised Biography*. HarperCollins, 1997.
Greenwalt, Dale E. *Remnants of Ancient Life. The New Science of Old Fossils*. Princeton University Press, 2022.
Humphrey, Louise and Stringer, Chris. *Our Human Story*. Natural History Museum, 2022.
Kenrick, Paul. *A History of Life in 50 Fossils*. Natural History Museum, 2020.
Kenrick, Paul and Davis, Paul. *Fossil Plants*. Natural History Museum, 2004.
Knoll, Andrew H. *Life on a Young Planet*. Princeton University Press, 2003.
Maisey, John G. *Discovering Fossil Fishes*. Henry Holt and Company, 1996.
McNamara, Ken. *Dragons' Teeth and Thunderstones. The Quest for the Meaning of Fossils*. Reaction Books Ltd., 2020.
Monks, Neale and Palmer, C. Phillip. *Ammonites*. Natural History Museum, 2003.
Naish, Darren. *Ancient Sea Reptiles*. Natural History Museum, 2022.
Naish, Darren and Barrett, Paul M. *Dinosaurs. How they Lived and Evolved*. Natural History Museum, 2019.
Savage, Robert J. G. and Long, Michael R. *Mammal Evolution*. Facts on File Publications, 1986.
Taylor, Paul D. and Lewis, David N. *Fossil Invertebrates*. Natural History Museum, 2019.
Taylor, Paul D. and O'Dea, Aaron. *A History of Life in 100 Fossils*. Natural History Museum, 2023.

Index

Page numbers in *italic* refer to illustration captions; those in **bold** refer to main subjects of boxed text.

Abereiddy Bay, Wales 122
acanthodians 149, *149*
Acanthostega 159, *159*
Acer 219
acritarchs 35, 234, 235
Acropora 109
Actinomma 229
Aegoceras 64
Aegyptopithecus 199, 200
Afrotherian placental mammals 187–90
Agassiz, Louis *148*
Agathoxylon 213, *213*
ages of fossils 8
    determination 20–1
agnathans *148*, 148–9
Agnostus 48
Akidograptus 122
Alasmotherium 194
algae 83, 109, 225, 245
    brown 218
    green 207, 218, 220
    red 35, *35*, 218, 220–1, *221*
    for blue-green algae *see* cyanobacteria
Algeria 169
Allonautilus 72
Alum Shales Formation, Sweden 48, *48*
Alvarez, Luis and Walter 26
Amaltheus 12

amber, fossils in *60*, 60–1, 167, 207, 217
ammolite 69
Ammonia 227
ammonites 18, 22, *27*, 63, 63–70
amniotes 163
amphibians, primitive 159–61
Amplexizaphrentis 110, 111
anapsids 163, 164
Anatolepis 148
Ancistrocrania 98
Ancyrocrinus 129, *129*
angiosperms 216–18
Anhanguera 173, *175*
Anning, Mary 154, *174*
Anomalocaris 41, *42*
Antarctica 139, 186, 212
antlers *196*, 197
apatite 30
Apex Chert, Australia 30
Apiocrinites 132, 133
Apiocystites 144
aptychus *68*, 69
Aptyxiella 90, 91
aragonite 9, 15, 16, 68–9, 74, 78, 79, 81, 83, 86, 91, 111
Araucaria 214
Archaea 31
Archaeocidaris 136
archaeocyaths *104*, 105–6
Archaeopteryx 58, 69, 179, *180*

Archean eon 29, 30
Archelon 165
Archimedes 116, *117*
archosaurs 163–81
Arctic 176
Arctica 78
Ardipithecus 202
Argentina 176, *163*, *184*, *191*
Arsinoitherium 189, 190
Art forms in nature (Haeckel) 229
Arthropleura 247
arthropods 45–61, 231
Asaphus 49, *49*
Asterias 127
asteroid impacts 26–7, 70, 235
asteroids (starfishes) 139, *140*
Asteroxylon 208
Astraeospongia 105, 106
Atlantic Ocean *23*, 27
Attenborough, Sir David 12
Australia 9, 10, 83, 120, 157, *181*, 186, 187, 212, 242
    New South Wales *7*, 186
    South Australia 35, *36*, *37*, 40, *104*, 119
    Victoria 186, *198*
    Western Australia 30, 31, 32–3, *33*
Australopithecus 202, *202*, 203
Austria *239*
avicularia 117, 118, *118*

bacteria 30, 31, 34, 37, 53, 82, 206, 240
Bangia 35
Bangiomorpha 35, *35*, 220
Barbados *230*
Barbaturex 167
barnacles *56*, 56–7, *57*
Barroisia 107, 108
Basilicus 46
Basilosaurus 199
bats 174, 183, *193*, 193–4
beetles 249
Belemnitella 75
belemnites 64, 73–5, *74*, *75*
Belgium *93*, *94*, *98*, 218
Bellerophon 89, *89*
Bembridge Marls, England 58
bennettitales 213, 214–15, *215*
'*Bereniced*' 118
Bering Sea 189
Bertie Group, Canada/USA 54
Bifrons Zone 70
bioluminescence 234
bipedalism 202, 245
birds 174, 179–81
bivalve molluscs 77–87, 95–6, *244*, *245*
Blaschka, Leopold and Rudolf 229
blastoids 142, 143, *144*, 145
Blue Lias Formation, UK 20

INDEX • 251

Bolsover Dragonfly 58
bone beds 249
borings and scrapings 242–5
Bothriolepis 149, 150, *150*, 158
brachiopods 17, 79, 94–101
Brachiosaurus 177
bracketing process 21
Bracklesham Group, England 86
Bradford Clay, England 133
Brasilia 66
Brazil 10, 175, 200, 231, *232*
breccia 13
British Museum, London, UK 152, 175
brittle stars *see* ophiuroids
Broken Hill, Zambia 203, *203*
Bromide Formation, Oklahoma, USA 95, 99
Brongniart, Alexandre 18
Brontosaurus 177
bryozoans 115–19
Buckland, William 248
building stones 11, 34, *34*, 52, 71, 72, 72–3, 77, 82, 83, *83*, 92, *110*, 111, *131*, 133, 220, 227–8
buoyancy 65, 67, 74, 75, 130
Burgess Shale, Canada 8, *29*, 39, 40, 41, *42*, 45, 54, 105
Buriolestes 177
burrows 37, 50, 79, 94, 238–42
butterflies 59

Calamites 211, *211*
Calcaribeyrichia 232
calcite 9, 15, 46, 49, 66, 68, 69, 78, 79, 80, 81, *90*, 91, 111, 124, 127, 225, *226*
calcium carbonate 32, 33, 38, 225
Callianassa 239
Calymene 45, 52, 52
Cambrian period 18
Cambrian Explosion 39–43
 palaeogeographical map 23
Campanile 12, 91
Canada
 Alberta 68, 69
 British Columbia 8, *29*, 39, 40, 41, *42*, 45, 54, 105
 Ellesmere Island 158, *158*
 Newfoundland 35, 37
 Nova Scotia 164
 Ontario 31, 34, 54, 80, 98, 112, *125*
 Quebec 150, 158
 Somerset Island 35
 St Joseph's Island 47
carapaces 165, *166*, 191

carbon dioxide, atmospheric 25, 26, 205, 225
Carboniferous Limestone 116
Carboniferous period
 coal forests 206, 209–12
 insects 60, 60–1, *61*
 palaeogeographical map 23
carcinization 56
cardinal fossula *110*, 111
carpoids 145
cartilaginous fishes 151–3
Carynella 143
Caulostrepsis 244
cave paintings, Palaeolithic 194
Caytonia 12, 215
Celliforma 241, *241*
Cenoceras 72
Cephalaspis 148, 148–9
cephalopods 63–75
ceratites 64, 70–1, *71*
Ceratites 71
Ceratodus 157, *157*
cetaceans 197–9
Chaetetes 106
chaetetids 106
chalcedony 206
Chalk, Cretaceous 15, *26*, 75, 82, 97, *98*, 108, *119*, *127*, 137, 138, 139, 225, *225*, 239
Chandler, Marjorie 217
Chaneya 219
Charnia 35, 36, 39
Charnwood Forest, England 35
Chatham Island, New Zealand 245
Cheiracanthus 149
chelonians 165–6
Chengjiang, China 40, *40*, 41, 45
chert 30, *31*, 32, 58, 208–9
Chesapecten 81, *81*
Chicxulub, Yucatan Peninsula, Mexico 27
Chile 198
China 34, *34*, 105, 130, 165, 175, 176, 201
 Guizhou 38–9, *39*
 Liaoning 177
 Lincheng 200
 Yunnan 40, *40*, 41, 45
Chinle Formation, Arizona, USA 213
chitons 93, 245
Chomatoseris 114, *115*
Chondrites 239
cirri 132, 133, *133*
clades 43
Cladophlebis 215
Clevosaurus 166, *167*

climate, changes in 22, 24–5, 205, 223
Cliona 243–4
Cloud, Preston 38
Cloudina 38, *38*
Cloughton Formation, England 215, *215*, 216, *216*
clubmosses 209, *235*
Clydagnathus 233
Clypeaster 9
Clypeus 135
coal balls 209
coal forests, Carboniferous 206, 209–12
Coal Measures, UK *61*
coccolithophores 15, 225
coccoliths 224–6, *226*
cockles 86
Coclichnus 247
coelacanths 157
Coelodonta 194, *194*
coiling 64, 65, 66, 68, 68, 88, 91
cold-bloodedness 163
collagen 119, 148
Colombia 168
coloniality 103, 110, 112, 114, 116, *117*, *118*, 118–19, *119*, 120, *121*, 122, *122*
Comura 51, *51*
concretions 14, 55, 61, *61*, 69, 206–7, 209
cones 207
conglomerate 13
conifers 213, *213*, 214
Conocardium 94
Conocoryphe see Elrathia
conodonts 223, 232–4
continental drift *23*, 24
conulariids 124–5, *125*
Conway Morris, Simon 41
Cooksonia 207–8, *208*
Copidozoum 118
coprolites 248–9
Corallian Group, England 114
Corallina 218
Coralline Crag, Suffolk, England 101, *101*, 119
corals 109–15
cordaitales 209, 211
Cordaites 211, *212*
Cornu 87
Cornulites 124
correlation of rocks 5, 21–2, 63, 70, 72, 120, 223, 233, 234
Cothurnocystis 145, *145*
Cottonwood Limestone, Kansas, USA 227–8
Crania 97
Craticularia 108

creodonts 193
Cretaceous period
 palaeogeographical map 23
 series 18
cribrimorphs 117, *118*
crinoids 89, *128*, 129–34
crocodylomorphs 168–9
Crocodylus 169
crustaceans 54–7
Cruziana 50, *246*, 247
cryptospores 234, 235
Cryptovaranoides 166–7
cuttlefish 75
Cuvier, Baron Georges 21, 224
cyanobacteria 31, 33, 34, 37, 205, 245
cycads 213, 214–15
Cyclothyris 101
Cylindroteuthis 75
cynodonts 184, *185*
Cynognathus 185, *185*
Cyprideis 232
Cyrtograptus 121
cystoids 142–3, *144*
Czechia 10, *121*

Dactylioceras 63, 64
Daimonelix 241, *241*, 242
Dapedium 154, *155*
Darwin, Charles 29, 56, 179, 216, 224
Darwinius 200
Darwinopterus 174
Deccan Traps 26, 27
deer, giant 11, *196*, 197
Dekayella 117
Deltatheridium 186
Deltoblastus 143, *144*, 145
Dendraster 139
Denmark *26*
Derbydene Marble, Derbyshire, England *131*
dermal scutes 169, *169*
diapsids 163, 164
diatomite 230, *230*
diatoms 228, 229–30
Dibunophyllum *110*, 111
Dickinsonia 35–6, *36*, 37, 39
Dictyonema 120, *121*
Didymograptus 122, *122*
Dimetrodon 164, *164*
Dimorphodon 174
Dinictis 192
dinoflagellates 234, *235*
Dinohippus 195
dinosaurs *163*, 175–9, 248, *248*
dinoturbation 248
Diplichnites 247
Diplocaulus 160, *161*

Diplodocus 177
Diprotodon 186, 187
DNA 43, 203
Dob's Linn, Scotland 122
dolomite 15
Doushantuo Formation, China 38–9, *39*
dragonflies 58, *58*, *59*
dragons 194
Dunkleosteus 149, 150, *150*

East Kirkton Quarry, West Lothian, Scotland 164
ecdysis *see* moulting and moults
Echinocardium 138
Echinocorys 135
echinoderms 127–45
echinoids *9*, 134–9
Echinosphaerites 143
Echinus 127
Edaphichnium 241
Edaphosaurus 164
Ediacaran fossils 35–9
Edmontosaurus 179
edrioasteroids 142, *143*
Egypt 190, *199*, 200, 228, *228*
Eldredgeops 50
Elrathia 12, *52*, 53
Emiliania 226, *226*
Encyonema 230
Endoceras 73
England
  Bristol *111*
  Cambridgeshire 155, *156*, 169
  Cornwall 98
  Cumbria 142, *143*
  Derbyshire 58, *110*, 131
  Devon *101*, 108
  Dorset 11, *84*, 90, 100, *130*, 141, *141*, *151*, 154, *155*, 169, *174*, 213, *213*, *235*, 239, 248
  Durham *110*, 111, *154*
  East Sussex 63, *118*, 176, 197
  Essex 87, 217
  Gloucestershire *88*, *135*, 220, *221*, *242*
  Hampshire *85*, *86*, 168, 217
  Herefordshire *121*, *208*
  Isle of Wight 58, *59*, *84*
  Kent 11, *57*, *108*, *153*, *181*, 195, *195*, 217, *225*, *244*
  Lancashire *89*
  Leicestershire 35
  London 72–3, *193*, *214*
  Mendip Hills *167*, 243
  Norfolk *75*, 202
  North Yorkshire 12, 63, *63*, 64, 70, *72*, 74, *97*, *131*, *168*,

169, 214, 215, *215*, 216, *216*, 239, *239*, 248
  Northamptonshire *137*
  Nottinghamshire *172*
  Oxfordshire 101, *107*, 108, 129, *137*, 176
  Shropshire *88*, *113*, *232*
  Somerset *2*, *69*, 70, *111*, *171*
  Staffordshire *61*
  Suffolk *77*, 87, 101, *101*, 119, *197*, 249
  Surrey *14*, *82*
  West Midlands *45*, *52*, *113*
  West Sussex *55*, 217
  Wiltshire 114, *115*, 128, *128*, *132*, 133, *133*
enrolment *50*, 52
Entobia 243, 244
environmental change, global 4–5, 22–7, 223
Eoconfuciusornis 179
Equisetum 211, 216
Equus 195–6
Eretmosaurus 172
Eryops 160, *160*
Ethiopia 202
Euarchontoglirean placental mammals 199–201
eukaryotes 29, 32, 34–5, 205
eurypterids 53–4
Eurypterus 54, *54*
Euspira 7
Eusthenopteron 158
Evelyn, John 217
evolution
  convergent 174, 187
  and fossils 4, 42–3, *224*
  theory of 29, 179

fake fossils *14*, *14*, 46
Faringdon Sponge Gravel, England *107*, 108
Favosites 112, *113*
Fayum Depression, Egypt 190, *199*, 200
feather stars 129, 133
feathers
  birds 179, *180*
  dinosaurs 177, *177*
Fenestella 117
fenestrules 116
ferns 209, 211, 213, 215, 235
ferrihydrite 206
finding fossils **10–11**
fishes 147–59
flint *14*, *14*, 108, 240, 243
flowers 217, 219
folklore names
  angels' money 228

beetroot stone 220–1, *221*
Brattingsborg pennies 98
Chedworth buns *135*
crystal apples 145
Delabole Butterfly 98
devil's toenail 80, *80*
Dudley Locust 52
Irish Elk *196*
Monster of Maastricht 167
Nutcracker Man 202
Osses 'Eds *84*
poundstones *135*
pundibs 101
screwstones 91
slaves' lentils 228
snakestones 63, *63*
St Cuthbert's Beads 131
starstones *115*, 129
thunderbolts 73, *75*
toadstones 154, *155*
fool's gold *see* pyrite
footprints and tracks 53, 175, 201–2, 245–6, *246*, 247–8
foraminifera 22, 94, 223, *223*, 224, 226–8, *228*
Fossil Grove, Glasgow, Scotland 210
fossilization
  of ancient microbial fossils 30
  of animals 8–9
  likelihood of 15–16
  of plants 9, 206–7
fossils
  and evolution 4, 42–3, **224**
  nature of 7
Foulden Maar, New Zealand 230, *230*
Fox Hills Formation, South Dakota, USA 27
Fractofusus 37
France 18, *65*, *83*, *84*, *107*, *115*, *118*, *133*, *240*
frogs *147*, 160
Frosterley Marble, Durham, England *110*, 111
frustules 230, *230*
Fucus 218, 237
fusain 206, 209, 217
fusulines 227
Futabasaurus 172

ganoid scales 154
Gastrochaenolites 244
gastroliths 173
gastropod molluscs 87–92
Gebel Zelten, Libya *192*
geological timescale 17–20

Germany 55, 71, 74, 154, *193*, 200
  Bundenbach 139, *140*
  Holzmaden 130, 171
  Ludwigsburg *157*
  Nattheim *103*, *114*
  Oeningen 218, *219*
  Rügen *119*
  Solnhofen 8, 58, 59, 68, 69, 179, *180*
  Westphalia 66
  Württemberg *165*
Gesner, Abraham 150
Gigantopithecus 200, 201
Gigantoproductus 100, *100*
Gilmerton Ironstone, Midlothian, Scotland *158*
Ginkgo 213, 215
glacial erratics 112
glauconite 21
Globigerina 226, *227*
Glossopteris 212, *212*
Glycymeris 86–7, *87*
Glyptodon 191, *191*
Glyptostroboxylon 206
Gnathichnus 245, *245*
Gobi Desert *178*
golden spikes 19
Gomphoceras 72, *73*
Gondwana 186, 191, 212
goniatites 64, 70, *71*, 71–2
Gotland, Sweden *107*, 112, *113*, *233*
Gould, Stephen J. 40
Graeophonus 61
Granton Shrimp Bed, Edinburgh, Scotland 233
graptolites 119–22
grasses 217
Great Ordovician Biodiversification Event 97
Great Oxidation Event 32
Greece 10
Greenland 40, 159, *159*
ground sloths, giant 190, *190*
Gryphaea 80, *80*
Gryphochron 93
Gunflint Chert, Canada *31*, 34
gymnosperms 213–14, 235

Hadean eon 29
Haeckel, Ernst 229
Halimeda 218
Hallucigenia 29, *40*, 41
Halysites 113
Hamelin Pool, Shark Bay, Australia 32–3, *33*
Hamites 65
heart urchins 137, *138*

INDEX • 253

Helicocoprus 249
Helicoprion 151, 152, *152*
Heliolites 112, *113*
Hell Creek Formation, Montana, USA 179
Helminthoidichnites 37, *37*
Hempstead Beds, England *92*
hesperornithiformes 179–80
Hibbertopterus 54
Hibolithes 74
Hildoceras 70
Hippopotamus 197, *197*
Hippurites 83
HMS *Challenger* expedition *223*
holdfasts 35, 129, *129*
Holectypus 137, *137*
holothurians 139, 141–2, *142*
holotypes *see* type specimens
Homarus 55
hominins *183*, 201–3, 245, *246*
Homo 10, 201, 203, *203*
'*Homo diluvii testis*' 218
Hoploscaphites 27
horses, evolution of 194–5
horsetails 211, *211*, 213, 215
Hou Xian-guang 40
House Range, Utah, USA *52*
human evolution 201–3
Hungary 206, 233
Hunsrück Slate, Germany 55, 139, *140*
hybodonts 151
Hybodus 151
Hydrochoerus 200
Hydrodamalis 189
hydroxylapatite 147, 151, 232
Hylaeosaurus 176
Hyracotherium 195, *195*

ichnology 237
ichthyosaurs 74–5, 169–71, *171*, 224
Ichthyosaurus 171
Ichthyostega 159
igneous rocks 13
Iguanodon 176, *176*
Imbrichnus 237
India 26, 198, 212, *212*
Indohyus 198
Indonesia *144*, 145, 203
Inoceramus 82
Insect Limestone, Isle of Wight, England *59*
insects 57–61
International Commission on Stratigraphy 19
Ipolytarnóc Fossil Forest, Hungary 206
Ireland 72, *196*, 197

iridescence *68*, 69, *177*; *see also* opalized fossils
iridium 26, *26*
ironstone formations 31
Isastrea 114
island dwarfism 189, 203
Isocrinus 130, 132, 133
Isotelus 47, 48
Italy 93, 156, *156*, *220*

Jaekelopterus 53
Jamaica 152, 190
Janjucetus 198, 199
Japan *172*, 233
Jefferson, Thomas 81
jellyfish 125
jet 214
jewellery, use of fossils in 52, 53, *60*, 63, 69, 154
Jiufotang Formation, China *177*
Jura Marble 11
Jura Mountains, France 18
Jurassic Park 17, 61
Jurassic period
  series 18
  stratigraphical stages 19

Kenya 197
kerogen 30, 57, 206
Kimberella 37
Koenigswald, Gustav von 201
Kokopellia 186
Koninckopora 220
Kumimanu 181
Kupferschiefer 154

La Brea, California, USA 192–3
Labidiaster 139
Laetoli, Tanzania 202, 245, *246*
Laevitrigonia 83
lagerstätten 8, 39, 40, 42, 45, 105, *193*
lamellibranchs *see* bivalve molluscs
Laminaria 218
lampshells *see* brachiopods
Lapworth, Charles 122
largest animals known from fossils
  ammonites 66
  amphibians 159–60
  arthropods 53
  belemnites 73, 74
  birds 180, 181
  brachiopods 100
  crinoids 134
  dinosaurs 177
  fishes 150, 152–3, 155, 158
  gastropods 91

insects 58
mammals 187, 188–9, 190, 191, *191*, 193, 197, 199
reptiles 164, 165, 167–8, 169, 170, 171
shells 82
trilobites 48
Latimeria 157
Laurasiatherian placental mammals 191–99
Leakey family 201, 245
leaves 207, *212*, 215, *219*
Leeds, Alfred 155
Leedsichthys 155, *156*
Leidy, Joseph *196*
Lepidodendron 207, 209–10, 210, *210*
Lepidophylloides 207
lepidosaurs 166
Lepidostrobus 207, 210
Lepidotes 154, *155*
Lhwyd, Edward 108
Libya *124*, *192*
limestones 15
limpets 88, *88*, 245
Lingula 94, 96, *97*
Linnaeus, Carl 12
lions 193
Liopleurodon 173
lissamphibians 160–1
lithification 8
Lithographus 247
Lithopsyche 59
Lithostrotion 111, 112
Littorina 91
lizards 163, 166, 167
lobe-finned bony fishes 157–9
Lobichnus 237
loboliths 129, 130
lobopodians 41
lobsters 56
London Clay, England 55, *55*, 85, 181, 195, *195*, 217, *244*
lungfish 157, *157*, 249
lycopods 209, *210*
Lyell, Charles 218

Mackie, William 208
magnesium 127
Malapa Cave system, South Africa *202*
mammals 183–203
mammoths *188*, 189
Mammut 188
Mammuthus 189
manganese dendrites 14
Mantell, Gideon 63
maps, palaeogeographical *23*, 24
Margulis, Lynn 34

marine plants 218–21
Marl Slate, England 154
Mars 4
Marston Marble, Somerset, England *69*, 70
marsupial mammals *186*, 186–7
Mason, Roger 35
mass extinctions *25*, 27
  end-Cretaceous (K-Pg) 18, 26, *27*, 58, 70, 75, 172, 179, 183, 235
  end-Permian 26, 58, 96, 116
  end Triassic 179
Massangis Limestone, France 133
Massetognathus 184, *184*
mastodons 188
Materpiscis 12
Mediterranean Sea 232
Meek, Fielding Bradford 12
Megaloceros *196*, 197
Megalosaurus 176
Meganeuropsis 58
Megatherium 190, *190*, 191
Megistotherium *192*, 193
Mellita 138
Merycoidodon *196*, 197
Mesohippus 195, *195*
Mesozoic Era
  palaeogeographical map *23*
Messel, Germany *193*, 200
Metaldetes 104
metamorphic rocks 13
meteorites 26
Mexico 27
Micraster 138, *138*
microbial fossils, ancient 30–2
Microconchus 124
microfossils 30, 223–35
Microraptor 177, *177*
moas 180, *247*
Modiolopsis 79, *80*
molecular clocks 34, 42, 43, 216
molluscs 77–94
Mongolia 176, *178*, 186
Monograptus 121
Monotrematum 186
monotreme mammals 186
Monte Bolca, Italy 156, *156*
Montlivaltia 114, *114*
Morganucodon 185, *185*
Morocco 51, *51*, 72, 200
Morrison, Jim 167
mosasaurs 167, *168*
moulds, internal *see* steinkerns
moulting and moults 45, 48, 53, 54, 231
mountain building 24
Mucrospirifer 99

multicellularity 29, 35
Multicostella 99
Murchison, Roderick 18
Muschelkalk 71
museum collections 11, 53, 152, 169
Muséum national d'Histoire naturelle, Paris, France 169
Myanmar 167
Myophorella 83, *84*
Mystacodon 199

Nahecaris 55
Nama Group, Namibia 38
names, scientific 12
Namibia 35, 38, *38*
nannofossils 224
Natica 91
Natural History Museum, London, UK 53, *196*
nautiloids 2, 11, 64, 70, 72, 72–3
Nautilus 72
Navajo Sandstone *248*
Neoceratodus 157
Neocrinus 132
neomorphism 9
neoselachians 151, 152
Neosolenopora 220–1, *221*
Neotrigonia 83
Neptunea 77, 91
Netherlands 167, 221
Neuropteris 205, 211
New Guinea 186
New Zealand 56, 57, 118–19, 166, 180, 181, 208, 230, *230*, 235, *245*, *247*
Niger 169
Niobrara Chalk, Kansas, USA 134
Northeast Red Formation, China 34
nudibranchs 87
Nummulites 228, *228*
Nyasasaurus 178, *178*
Nymphaster 139, *140*
Nypa 217, *218*

ocean formation *23*, 24
oceanic crust 25
octopuses 64, 75, 86, 244
Odontaspis 153
Odontochelys 165
Oichnus 244–5, *245*
Öland Limestone, Sweden 11, *72*
Old Red Sandstone 148, *148*
oldest fossils 29–43, 57
Olduvai Gorge, Tanzania *183*, 201, 202

Olivellites 237
Oneirophantites 142
Opabinia 41
opalized fossils *7*, 9
Ophiomorpha 238
ophiuroids 139, 141, *141*
Oppelia 68
Oriostoma 88, 89
ornithischians 176
Orthoceras Limestone 73
Osborn, Henry Fairfield 12
ossicles 127, 128, 129, 131, 139
ostracods 223, 231–2
Otodus 152, *153*
Ototara Limestone, New Zealand 118–19
Owen, Richard 175, *186*
Oxford Clay, England *156*
oxygen, atmospheric 31–2, 58, 205, 209
Oxyopes 60
oysters 80, *80*, *123*, *245*

Pakicetus 198, *198*
Pakistan *198*
Palaeocaster 242
Palaeochiropteryx *193*
Palaeocoma 141, *141*
palaeogeographical maps *23*, 24
palaeognaths 180
Palaeolagus 200
palaeoniscum 154, *154*
Palaeoniscum 154, *154*
Palaestella 140
Paleodictyon 240, *240*
Paleryx 168
pallial line 78, *79*, 86
palynomorphs 223, 234–5
Panama *118*
Pander, Heinz Christian 232
Pangaea 209
Panopea 79
Panthera 193
Paraconularia 125
Paranthropus *183*, 202
Parapassaloteuthis 74
Parapuzosia 66
Parkinsonia 66
partial mortality 103
Patagonia 186, 190, *190*
Patagornis 181
Patagotitan 163, *177*
Pattersoncypris 231
Pearce, Joseph Chaning 171
pearly nautilus 64, 67
peat 209, 211
pedicle 95, 96, 97
pelecypods *see* bivalve molluscs
Pelmatopora 118
Pelophylax 147, 160

pelycosaurs 164, *164*, 184
penguins 181
Pentacrinites 130, *130*, 132
pentamerism 127
Permian period 18
permits for fossil collecting 10
Peru 193, 199
Petrified Forest National Park, Arizona, USA 213, *213*
Pezosiren 190
Phalaphium 60
Pholadomya 85, *85*
phorusrhacids 180–1
phosphate, agricultural 249
phosphatization 9, 160, 231
photosynthesis 31, 34, 109, 205, 207, 227, 230
Pierre Shale, South Dakota, USA 165
pinnules 129, 132
pipe rock 238
placental mammals 187–201
Placenticeras 68, 69
placoderms 149–50, *150*
Plagiophthalmosuchus 168, 169
plankton blooms 234
plants 205–21
plate tectonics 22, 24
Platecarpus 168
Platyceramus 82
platyceratids 89
Platychonia 107
plesiosaurs 169, 171–3
Pleuromya 85
pleurosaurids 166
Pleurotomaria 90, 91
Pliothyrina 101, *101*
Plot, Robert 129
Poland 249
Poleumita see Oriostoma
pollen 234, 235
Polycotylus 173
polymorphism 117
polyplacophorans 93, *93*
Portland Stone, England 11, 83, 91, 220
Portugal 83
Posidonia Shale, Germany 171
Precambrian 29, 35
primates 200–3
Procoptodon 187
productids 98, 100, *100*
Proganochelys 165, *165*
prokaryotes 31
Prophaeton 181
Proterozoic eon 29
Protoceratops 178
Protocupressinoxylon 213, 214
Prototaxites 212

Pseudodiplocoenia 114, *115*
pseudofossils **14**
Pteranodon 173
Pteraspis 149
pteridosperms 205, 209, 211, 213, 215
pterosaurs 173–5
punctae 100
Purbeck Marble, England 11, 92
pycnofibres 175
Pygaster 137, *137*
pyrite 9, 48, 69, 139, *140*, 206, 217
pyroclastics 15

rabbits 200
radiodontids 29, 41, *42*
radiolarians 228–9, *229*
radiometric dating 18, 20–1
Radulichnus 245
Rafinesquina 98, *99*
rangeomorphs 36–7
Raphidonema *107*, 108
ray-finned bony fishes 153–6
rays 151, 152, *153*
Red Crag Formation, England 77, 87, 91, *197*
reefs 106, 107, 109, 112, 114, 156, 220
Reid, Eleanor 217
reproduction
    asexual 228
    egg-laying 175, 177–8, *178*, 185, 186
    live births 171, 173
    sexual 35, 228
reptiles 163, 164–75
respiration 34, 45, 54, 143, 157, 159
Rhabdinopora 120
Rhabdopleura 120
Rhincodon 155
Rhizodus 158, *158*
rhodoliths 220, *220*
rhodophytes *see* algae: red
rhynchocephalians 166, *167*
Rhynie Chert 58, *208*, 208–9
Rigaudella 235
rocks
    correlation 5, 21–2, 63, 70, 72, 120, 223, 233, 234
    dating 20–1
    types 13, 15
rodents 183, 199–200
roots 207, *210*, 216
rostroconchs 94, *94*
rudists 82, 83
Russia 18, 26, 32, 35, *49*, 73, *106*, 189, 194
Rutherford, Ernest 20

sabretooths 192, *192*
sand dollars 137, *138*, 138–9
Santana Formation, Brazil 231
Sarcosuchus 169
saurischians 176–7
Savage, Bob *192*
scallops 81
scaphopods *93*, 93–4
Scheuchzer, Johann Jakob 218
Schopf, Bill 30
sclerotic rings 171, *171*
Scotland 238
   Aberdeenshire 58, 208–9
   Ayrshire 145, 229
   Banffshire *149*
   Dumfries and Galloway 122
   Edinburgh *158*
   Firth of Forth 233
   Forfar *148*
   Glasgow *125*, 210
   Isle of Skye 73
   West Lothian 164
Scott, Robert Falcon 212
Scyphocrinites 129
sea biscuits 137
sea cows 189–90
sea cucumbers *see* holothurians
sea levels 22, 24, 25, *25*
sea lilies *see* crinoids
sea scorpions *see* eurypterids
sea urchins *see* echinoids
seagrasses 221
seaweeds 34, 218
Sedgwick, Adam 18
sedimentary rocks 13
seed ferns *205*, 209, 211, 213, 215
seeds *219*
Seirocrinus 130
Selaginella 235
Septastrea 114, 114–15
Seriola 156
sexual dimorphism 68, 179
sexual reproduction 35, 228
sharks *151*, 151–3, *153*, 249
shell beds 77
shelled reptiles 165–6
shells 77–101
Siberian Traps 26, 27
Sifrhippus 195
Sigillaria 210
Silesaurus 249
silica 229
Siphonia 108, *108*
Siphonodendron 111, 112
sirenians 189–90
Skolithos 238, 239, 243
Sloane, Hans 152, *153*
Smilodon 192, *192*, 193

Smith, William 21, 100–1
Smithsonian Institution, Washington DC, USA 39
snakes 163, 167–8
soft part fossilization 8, 37, 38, 40, 42, 61, 74, 123, 148, 160, 161, 170, 174, 180, 231, 233–4
Solnhofen Limestone, Germany 8, 58, *59*, *68*, 69, 179, *180*
Soom Shale, South Africa 234
South Africa 157, *185*, 189, 201, *202*, 234
Spain *9*, 142, *142*, *147*, 160, 214, 228, 240, *241*, *246*
Sphenodon 166
sphenosuchians 169
spicules 105, *105*, 107, 141
spiders 57, 60, *60*, 61, *61*
spines *127*, *136*, 137, 149, 151, *151*, 228
Spirogyra 207
Spirorbis 124
Spirula 74
sponges 104–8, 243–4
spore capsules 208, *208*
spores 234, 235, *235*
spreiten 240
Sprigg, Reg 35
squids 75
St Cuthbert 131
St Hilda 63, 70
Stalticodiscus 143
Starfish Bed, Dorset, England 141
starfishes *see* asteroids
state fossils, US 81, 199
Stegosaurus 176
steinkerns *9*, 9, 69, 79, 80, *84*, 85, *90*, 91
Stenopterygius 170, 171
Stethacanthus 151
Stevns Klint, Denmark *26*
Stigmaria 207, 210, *210*
Štramberk Limestone, Czechia *10*
Stramentum 57
strata 13
stratigraphical stages 19
Strelley Pool Formation, Australia 31
stromatolites 30, 32–4, 37, 213, *214*
stromatoporoids 106–7
Stromatoveris 36
Stylemys 166, *166*
subduction 24
subfossils 8, 180
suture lines 71, *71*
Suzuki, Tadashi 172

Sweden 11, 48, *48*, 72, 73, *98*, 107, 112, *113*, 143, *233*
Switzerland 18
symbiogenesis 34
symbionts 82, 83, 109, 205, 226, 227
Symmetrocapulus 88
symmetry 127, 137
Symplocos 217
synapsids 163, 164, 165

tadpoles 160, *161*
Tanzania 178, *178*, *183*, 201, 202, 245, *246*
taphonomy 16
teeth
   dinosaurs *176*
   fishes 147, 151, 152, *152*, *153*, 154, *155*, *158*
   mammals 183, 184, *184*, 185, *185*, 186, 187, 188, *188*, 192, *192*, 195, *195*, *197*, 199, *200*, 201, 202
   reptiles 164, 166, 167, *168*, 169, 173, *173*
Teinolophos 186
Telicomys 200
temnospondyls 159–60, *160*, 164
temperatures
   global 24–5, *25*
   seawater 227
Tentaculites 124, *124*
terebratulids 100
Teredolites 244, *244*
terror birds 180, *181*
Tethys Ocean 82
tetrapodomorphs 158, *158*
Thalassinoides 239, *239*, 240
thalattosuchians 169
Thecosmilia 103, 114
therapsids 184–5
thermoregulation 165
thrombolites 33
Thylacinus 187
Thylacoleo 187
Thylacosmilus 187
Tianzhushania 39
Tiktaalik 158, 158–9
time-averaging 16
Titanoboa 167
toothplates 157, *157*
tortoises 165, *166*
trace fossils 7, *237*, 237–49
tracks and footprints 53, 175, 201–2, 245–6, *246*, 247–8
trails 50, 246–7
Traumatocrinus 130
tree trunks 206, 207, *214*

trees, earliest 212
Triceratops 176, 179
Tridacna 78, 83
Trigonia 84
trigoniids 83, *84*
trilobites 18, *45*, 45–53, 224, *246*, 247
Trinucleus 51
Trueman, Arthur 80
Trypanites 242, 243
tuataras 166
tubeworms 124
Tupus 58
Turanophlebia 59
Turkey *161*
Turrilites 65
tusk shells *93*, 93–4
Tylocidaris 127, 137
tympanic bullae *197*, 198
type localities 20
type specimens 12, *82*, *158*, *181*
Tyrannosaurus 12, *176*, 177, 178, 179

Uinta Mountains, USA 134
Uintacrinus 134, *134*
UK *see* England, Scotland *and* Wales
ungulates 194–8
uniformitarianism 205, 238
Unio 85, *85*
unionids 85
United States
   Alabama 199
   Arizona 213, *213*
   California 192–3
   Colorado 82
   Florida 86, *114*, 115
   Illinois 206
   Indiana *129*
   Iowa *117*
   Kansas 58, 134, *134*,173, 228
   Kentucky *143*
   Maryland 152, *153*
   Mississippi 199
   Missouri *188*
   Montana 179
   Nebraska *166*, *195*, 197
   Nevada *152*
   New York State *54*, 54, 73, *123*
   North Carolina *123*
   Ohio *99*, 116, *117*, *150*
   Oklahoma *95*, *99*
   Pennsylvania *212*
   South Carolina *138*
   South Dakota *27*, 165, *196*, 196–7, 241, *241*
   Tennessee *105*

Texas *136*, *160*, 164
Utah 11, *52*, 53, 173, 186, 248
Virginia 79, 81, *81*, *153*

Velociraptor 177
Venericor 86, *86*
vertebrae 147, 168, 172
vertebrates 147–203
Victoria, Queen 20, 214
Virgin Formation, Utah, USA 11
Viviparus 92, *92*
volcanic eruptions 26, 27

Walcott, Charles Doolittle 39–40, 41
Wales 18, *51*, 112, *122*, 185
   Flintshire *117*
   Glamorgan *185*
   Pembrokeshire 122
Wallace, Alfred Russel 224
warm-bloodedness 170, 173, 175
Weald, England 85
Wealden Marble, England 92, *92*
Wenlock Limestone, UK 16, *17*, 45, 52, 89, *113*, 144
Western Sahara *50*

Westlothiana 164
whales *197*, 197–8, *198*
Wheeler Formation, Utah, USA 53
White Sea 35
wings, insect
   evolution 58
   use in palaeoentomology 60
wingspans 175, *175*, *180*
wood, fossil 206, 213, 217
woolly rhinoceros 188, 194, *194*
World Heritage Sites 33
'worms' and conical tubes 94, 123–5

Xenarthran placental mammals 190–1

Yoredale Shales, England *131*

Zambia 203, *203*
Zamites 215
Zigzagopora 99
zircon 21
Zoophycos 240, *240*
zooxanthellae 109

Acknowledgements

For expert comments on individual chapters, or parts of chapters, I am grateful to Katie Collins, Tim Ewin, Richard Fortey, Jerry Hooker, Paul Kenrick, Christian Klug, Zerina Johanson, Emanuela Di Martino, Susie Maidment, Giles Miller, Leigh Anne Riedman, Chris Stringer, Stephen Stukins, Richard Twitchett, Andrea Waeschenbach and Mark Wilson. My wife Patricia Taylor and friend Tim de Ferrars kindly read the entire text. Several colleagues gave freely of their time when locating specimens for photography, including Neil Adams, Emma Bernard, Katie Collins, Jill Darrell, Rob Day, Tim Ewin, Peta Hayes, Jerry Hooker, Richie Howard, Zoe Hughes, Rachel Ives and Marc Jones. The constructive comments of two anonymous reviewers are acknowledged.

Picture credits

pp.10, 11, 14, 26, 34, 56, 72 (top), 83 (left), 92 (right), 95, 110 (right), 131 (bottom), 142, 219, 220, 239 (top), 245 (left) ©P.D. Taylor; p.28 ©John Sibbick/Science Photo Library; p.33 ©D.P. Gordon; p.35 ©N.J. Butterfield; p.36 ©J. Gehling; p.38 ©R. Wood; p.42 ©The Trustees of the Natural History Museum, London, specimen from Royal Ontario Museum; p.185 (top) ©Pamela Gill; p.185 (bottom) ©Christine Strullu-Derrien; p.214 ©Paul Kenrick; p.230 (bottom) ©Image by Uwe Kaulfuss, *Fossil Treasures of Foulden Maar: A window into Miocene Zealandia*. Otago University Press, 2022, p. 50; p.241 ©UNL Office of Research; p.248 ©Susannah Maindment.

Unless otherwise stated images copyright of The Trustees of the Natural History Museum, London.
Every effort has been made to contact and accurately credit all copyright holders. If we have been unsuccessful, we apologise and welcome correction for future editions and reprints.